# ANAEROBIC OXIDATION OF METHANE COUPLED TO THE REDUCTION OF DIFFERENT SULFUR COMPOUNDS AS ELECTRON ACCEPTORS IN BIOREACTORS

Chiara Cassarini

Joint PhD degree in Environmental Technology

UNIVERSITÉ ━━━
━PARIS-EST

Docteur de l'Université Paris-Est
Spécialité : Science et Technique de l'Environnement

Dottore di Ricerca in Tecnologie Ambientali

UNESCO-IHE
Institute for Water Education

Degree of Doctor in Environmental Technology

## Tesi di Dottorato – Thèse – PhD thesis

Chiara Cassarini

## Anaerobic oxidation of methane coupled to the reduction of different sulfur compounds as electron acceptors in bioreactors

Thesis defended on June 28th, 2017

In front of the PhD committee

| | |
|---|---|
| Prof. Dr. Erkan Sahinkaya | Reviewer |
| Prof. Dr. Artin Hatzikioseyian | Reviewer |
| Prof. Dr. Piet N. L. Lens | Promotor |
| Prof. Giovanni Esposito | Co-Promotor |
| Dr. Eldon R. Rene | Co-Promotor |
| Dr. Eric D. van Hullebusch | Examiner |

European Commission
ERASMUS
MUNDUS

Erasmus Joint doctorate programme in Environmental Technology for Contaminated Solids, Soils and Sediments (ETeCoS³)

**Thesis Committee**

**Thesis Promoter**

Prof. Dr. Piet N. L. Lens
Professor of Environmental Biotechnology
UNESCO-IHE, Institute for Water Education
Delft, the Netherlands

**Thesis Co-Promoters**

Dr. Giovanni Esposito
Associate Professor of Sanitary and Environmental Engineering
University of Cassino and Southern Lazio, Cassino, Italy

Dr. Eldon R. Rene
UNESCO-IHE, Institute for Water Education
Delft, the Netherlands

**Examiner**
Dr. Eric D. van Hullebusch, PhD
Associate Professor of Biogeochemistry
University of Paris-Est, Institut Francilien des Sciences Appliqueées
Champs sur Marne, France

**Reviewers**
Prof. Dr. Erkan Sahinkaya
Istanbul Medeniyet University
Bioengineering Department
Goztepe, Istanbul, Turkey

Prof. Dr. Artin Hatzikioseyian
National Technical University of Athens
School of Mining and Metallurgical Engineering
Zografou, Greece

This research was conducted under the auspices of the Erasmus Mundus Joint Doctorate Environmental Technologies for Contaminated Solids, Soils, and Sediments (ETeCoS$^3$) and The Netherlands Research School for the Socio-Economic and Natural Sciences of the Environment (SENSE).

CRC Press/Balkema is an imprint of the Taylor & Francis Group, an informal business

Published by:
CRC Press/Balkema
Schipholweg 107C, 2316 XC, Leiden, the Netherlands
Pub.NL@taylorandfrancis.com
www.crcpress.com – www.taylorandfrancis.com
ISBN  978-1-138-32913-3 (Taylor & Francis Group)

**Table of contents**                                                          **Page no.**

## Acknowledgements

I would first like to thank the Erasmus Mundus Joint Doctorate ETeCoS[3] program (FPA n°2010-0009) for providing the financial support to carry out this PhD research. I would like to thank the Institution and Universities that hosted me along these years of research: UNESCO-IHE, Institute for Water Education (Delft, the Netherlands), Shanghai Jiao Tong University (China), the Helmholtz-Centre for Environmental Research - UFZ (Leipzig, Germany) and the University of Cassino and Southern Lazio (Italy). It has been a great learning experience working in different labs and environments in four diverse countries.

I would like to express my sincere gratitude to my supervisor and PhD promotor, Prof. Piet N. L. Lens, for his priceless suggestions, inspirations, constant supervision during the progress of my research and valuable supports throughout the research and this thesis.

Special thanks to my PhD co-promoter, Dr. Giovanni Esposito for his supports and inspirations during this research. Further, I would like to acknowledge Dr. Eric D. van Hullebusch, for his support during entire PhD program.

I cannot be more thankful to my mentor and my thesis co-promoter Dr. Eldon R. Rene for his constant help, his suggestions during the entire research and refining the manuscripts and the thesis. I would like to thank all my mentors in the different institutes and universities I have been during the PhD research: Dr. Graciela Gonzalez-Gil for the initial support in the experiment set-up, Dr. Yu Zhang for the microbiological analysis and high pressure experiments, Dr. Carsten Vogt for his support with the isotope analysis and Dr. Niculina Musat for her support during the microscopy analysis.

I would like to acknowledge Dr. Jack Vossenberg for his help and advices in the microbial analysis. Further, I would like to thank Dr. Filip Meysman and his team at the laboratory of Royal Netherlands Institute of Sea Research NIOZ, Yerseke, the Netherlands for providing the Lake Grevelingen sediments and fruitful discussions about the lake biochemistry. I would like to acknowledge Prof. Caroline P. Slomp and Dr. Matthias Egger for the collaborative work in the marine Lake Grevelingen.

I would like to thank two master students, Zita Naangmenyele and Dessie Bitew for their contribution in this research work. Moreover I would like to thanks entire laboratory team of UNESCO-IHE for their support in the laboratory. Thanks to all the colleagues from the State Key Laboratory of Microbial Metabolism/ Laboratory of Microbial Oceanography, Shanghai Jiao Tong University. A special thank to all the staff and colleagues at the Department of Isotope Biogeochemistry and at the Centre for Chemical Microscopy (ProVIS) at the UFZ (Leipzig, Germany), especially Dr. Hryhoriy Stryhanyuk, Dr. Ivonne Nijenhuis, Prof. Hans-H. Richnow.

A very special thank to my friend and colleague, Dr. Susma Bhattarai for all the help and suggestions you gave me, for all the hard work we did together, for the moral support and friendship. Working with you it has been a long and enjoyable journey. Thanks to all the friends and colleagues with whom I shared time during this reasearch in the Netherlands, China,

Germany and Italy. Thanks to Iosif, Joy, Arda, Shrutika, Dr. Jian Ding, Yan Wan Kai, Xiaoxia Liu, Paolo, Gabriele, Francesco and Kirki for their help in different stages of this research.

Finally, I would like to thank my parents who always gave me support and comfort and my entire family. A special thank to Rick de Groot who gave me support and has always been by my side during this entire PhD journey and Dr. Tom Bosma who encouraged me to do this PhD research. The entire work would not be possible without them.

**Summary**

Large amounts of methane are generated in marine sediments, but the emission to the atmosphere of this important greenhouse gas is partly controlled by anaerobic oxidation of methane coupled to sulfate reduction (AOM-SR). AOM-SR is mediated by anaerobic methanotrophs (ANME) and sulfate reducing bacteria (SRB). AOM-SR is not only regulating the methane cycle but it can also be applied for the desulfurization of industrial wastewater at the expense of methane as carbon source. However, it has been difficult to control and fully understand this process, mainly due to the slow growing nature of ANME. This research investigated new approaches to control AOM-SR and enrich ANME and SRB with the final purpose of designing a suitable bioreactor for AOM-SR at ambient pressure and temperature. This was achieved by studying the effect of (i) pressure and of (ii) the use of different sulfur compounds as electron acceptors on AOM, (iii) characterizing the microbial community and (iv) identifying the factors controlling the growth of ANME and SRB.

Theoretically, elevated methane partial pressures favor AOM-SR, as more methane will be dissolved and bioavailable. The first approach involved the incubation of a shallow marine sediment (marine Lake Grevelingen) under pressure gradients. Surprisingly, the highest AOM-SR activity was obtained at low pressure (0.45 MPa), showing that the active ANME preferred scarce methane availability over high pressure (10, 20, 40 MPa). Interestingly, also the abundance and structure of the different type of ANME and SRB were steered by pressure.

Further, microorganisms from anaerobic methane oxidizing sediments were enriched with methane gas as the substrate in biotrickling filters (BTF) at ambient conditions. Alternative sulfur compounds (sulfate, thiosulfate and elemental sulfur) were used as electron acceptors. When thiosulfate was used as electron acceptor, its disproportionation to sulfate and sulfide was the dominating sulfur conversion, but also the highest AOM-SR rates were registered in this BTF. Therefore, AOM can be directly coupled to the reduction of thiosulfate, or to the reduction of sulfate produced by thiosulfate disproportionation. Moreover, the use of thiosulfate triggered the enrichment of SRB. Differently, the highest enrichment of ANME was obtained when only sulfate was used as electron acceptor.

In a BTF with sulfate as electron acceptor, both ANME and SRB were enriched from marine sediment and the carbon fluxes within the enriched microorganisms were studied through fluorescence in-situ hybridization-nanometer scale secondary ion mass spectrometry (FISH-NanoSIMS). Preliminary results showed the uptake of methane by a specific group of SRB.

ANME and SRB adapted to deep sediment conditions were enriched in a BTF at ambient pressure and temperature. The BTF is a suitable bioreactor for the enrichment of slow growing microorganisms. Moreover, thiosulfate and sulfate can be used strategically as as electron acceptors to activate the sediment and enrich the SRB and ANME population to obtain high AOM-SR and faster ANME and SRB growth rates for future environmental applications.

## Samenvatting

Grote hoeveelheden methaan worden gegenereerd in mariene sedimenten, maar de uitstoot in de atmosfeer van dit belangrijke broeikasgas wordt gedeeltelijk beheerst door anaërobe oxidatie van methaan gekoppeld aan sulfaat reductie (AOM-SR). AOM-SR wordt gemedieerd door anaerobe methanotrofen (ANME) en sulfaatreducerende bacteriën (SRB). AOM-SR reguleert niet alleen de methaancyclus, maar kan ook worden toegepast voor de ontzwaveling van industrieel afvalwater ten koste van methaan als koolstofbron. Het was echter moeilijk om dit proces te beheersen en volledig te begrijpen, voornamelijk vanwege de traaggroeiende aard van ANME.

Dit onderzoek onderzocht nieuwe benaderingen om AOM-SR te controleren en ANME en SRB te verrijken met als uiteindelijk doel het ontwerpen van een geschikte bioreactor voor AOM-SR bij omgevingsdruk en -temperatuur. Dit werd bereikt door het effect van (i) druk en (ii) het gebruik van verschillende zwavelverbindingen als elektronenacceptoren op AOM te bestuderen, (iii) de microbiële gemeenschap te karakteriseren en (iv) de factoren te identificeren die de groei van ANME en SRB beheersen.

Theoretisch, verhoogde methaan partiële druk gunstig AOM-SR, omdat meer methaan zal worden opgelost en biologisch beschikbaar zal zijn. De eerste benadering betrof de incubatie van een kustsediment (Grevelingenmeer) onder drukgradiënten. Verrassenderwijs werd de hoogste AOM-SR-activiteit verkregen bij lage druk (0,45 MPa), wat aantoont dat de actieve ANME de voorkeur gaf aan een schaarse beschikbaarheid van methaan boven hoge druk (10, 20, 40 MPa). Interessant is dat ook de overvloed en structuur van het verschillende type ANME en SRB werden gestuurd door druk.

Verder werden micro-organismen uit anaerobe methaanoxiderende sedimenten verrijkt met methaangas als het substraat in biotricklingfilters (BTF) bij omgevingscondities. Alternatieve zwavelverbindingen (sulfaat, thiosulfaat en elementaire zwavel) werden gebruikt als elektronenacceptoren. Wanneer thiosulfaat als elektronenacceptor werd gebruikt, was de disproportionering ervan met sulfaat en sulfide de dominerende zwavelconversie, maar ook de hoogste AOM-SR-snelheden werden in deze BTF geregistreerd. Daarom kan AOM direct worden gekoppeld aan de reductie van thiosulfaat, of worden gereduceerd door sulfaat geproduceerd door thiosulfaat disproportionering. Bovendien triggert het gebruik van thiosulfaat de verrijking of SRB. Anders werd de hoogste verrijking van ANME verkregen wanneer alleen sulfaat als elektronenacceptor werd gebruikt.

In een BTF met sulfaat als elektronenacceptor werden zowel ANME als SRB verrijkt met mariene sedimenten en de koolstoffluxen in de verrijkte micro-organismen werden bestudeerd door middel van fluorescentie in-situ hybridisatie - nanometer schaal secundaire ionen massaspectrometrie (FISH-NanoSIMS). Voorlopige resultaten toonden de opname van methaan door een specifieke groep SRB.

ANME en SRB aangepast aan diepe sedimentomstandigheden werden verrijkt in een BTF bij omgevingsdruk en temperatuur. De BTF is een geschikte bioreactor voor de verrijking van langzaam groeiende micro-organismen. Bovendien kunnen thiosulfaat en sulfaat strategisch worden gebruikt als elektronenacceptoren om het sediment te activeren en de SRB- en ANME-populatie te verrijken om hogere AOM-SR en snellere ANME- en SRB-groeisnelheden voor toekomstige toepassingen te verkrijgen.

**Sommario**

Grandi quantità di metano vengono generate nei sedimenti marini, ma l'emissione atmosferica di questo importante gas serra è parzialmente controllata dall'ossidazione anaerobica del metano accoppiata alla solfato riduzione (OAM-SR). OAM-SR è mediato da metanotrofi anaerobici (ANME) e da batteri solfato riduttori (SRB). OAM-SR non solo regola il ciclo del metano ma può anche essere applicato alla desolforazione delle acque reflue industriali usando il metano come fonte di carbonio. Tuttavia, è difficile controllare e comprendere appieno questo processo, principalmente a causa della lenta crescita degli ANME. In questa ricerca vengono considerate nuove strategie per controllare l'OAM-SR e arricchire gli ANME e gli SRB al fine di progettare un bioreattore adatto per l'OAM-SR a pressione e temperatura ambiente. Ciò è stato ottenuto studiando l'effetto della (i) pressione e (ii) l'uso di diversi composti di zolfo come accettori di elettroni su OAM, (iii) caratterizzando la comunità microbica e (iv) identificando i fattori che controllano la crescita degli ANME e SRB.

Teoricamente, elevate pressioni parziali del metano favoriscono l'OAM-SR, poiché più metano viene disciolto. Il primo approccio prevedeva l'incubazione di sedimenti costieri di un lago salmastro (lago Grevelingen) a diversi gradienti di pressione. La più alta attività di OAM-SR è stata ottenuta a bassa pressione (0,45 MPa), dimostrando che gli ANME attivi preferivano scarsa disponibilità di metano rispetto all' alta pressione (10, 20, 40 MPa). Da notare che la pressione ha anche influenzato l'abbondanza e la struttura dei diversi tipi di ANME e SRB.

Inoltre, i microrganismi provenienti da sedimenti anaerobici, capaci di ossidare il metano, sono stati arricchiti usando metano come substrato in letti percolatori (LP) a condizioni ambientali. Diversi composti dello zolfo (solfato, tiosolfato e zolfo elementare) sono stati usati come accettori di elettroni. Quando il tiosolfatoè stato usato come accettore di elettroni, la sua conversione dominante era il disproporzionamento a solfato e solfuro, ma anche i tassi più alti di OAM-SR sono stati registrati in questo LP. Pertanto, l'OAM può essere direttamente accoppiato alla riduzione del tiosolfato o alla riduzione del solfato prodotto dal disproporzionamento del tiosolfato. Inoltre, l'uso di tiosolfato ha indotto all'arricchimento degli SRB. Diversamente, il più alto arricchimento di ANME è stato ottenuto quando il solfato è stato usato come solo accettore di elettroni. .

Nel LP con solfato come accettore di elettroni, sia gli ANME che gli SRB sono stati arricchiti da sedimenti marini e gli scambi di carbonio tra i microrganismi arricchiti sono stati studiati mediante ibridazione fluorescente *in situ* e spettrometria di massa di ioni secondari in scala nanometrica (FISH-NanoSIMS). I risultati preliminari hanno mostrato l'assorbimento di metano da parte di un gruppo specifico di SRB.

ANME e SRB adattati a sedimenti profondi sono stati arricchiti in un LP a pressione e temperatura ambiente. LP è un bioreattore adatto per l'arricchimento di microrganismi a crescita lenta. Inoltre, il tiosolfato e il solfato possono essere utilizzati strategicamente come accettori di elettroni per attivare il sedimento e arricchire più rapidamente la popolazione di SRB e ANME per applicazioni future.

## Résumé

De grandes quantités de méthane sont générées dans les sédiments marins, mais l'émission dans l'atmosphère de cet important gaz à effet de serre est partiellement contrôlée par l'oxydation anaérobie du méthane couplée à la réduction du sulfate (AOM-SR). AOM-SR est médiée par les méthanotrophes anaérobies (ANME) et sulfate bactéries réductrices (SRB). AOM-SR ne régule pas seulement le cycle du méthane, mais il peut également être appliqué pour la désulfuration des eaux usées industrielles en utilisant du méthane comme source de carbone. Cependant, il est difficile de contrôler et de comprendre pleinement ce processus, principalement en raison de la lente croissance de l'ANME.

Cette recherche a étudié de nouvelles approches pour contrôler AOM-SR et enrichir ANME et SRB afin de concevoir un bioréacteur approprié pour AOM-SR à pression et température ambiantes. Ceci a été réalisé en étudiant l'effet de (i) la pression et (ii) l'utilisation de différents composés de soufre comme accepteurs d'électrons sur OAM, (iii) la caractérisation de la biomasse enrichie et (iv) l'identification des facteurs contrôlant la croissance de ANME et SRB.

Théoriquement, des pressions partielles élevées de méthane favorisent AOM-SR, car plus de méthane est dissous. La première approche impliquait l'incubation des côtiers d'un lac saumâtre (lac Grevelingen) à différents gradients de pression. L'activité la plus élevée de AOM-SR a été obtenue à basse pression (0,45 MPa), démontrant que l'ANME active préférait la la faible disponibilité du méthane à la haute pression (10, 20, 40 MPa). Il est intéressant de noter que la pression a également influencé l'abondance et la structure des différents types d'ANME et de SRB.

De plus, les micro-organismes provenant des sédiments anaérobies, capables d'oxyder le méthane, ont été enrichis avec du méthane comme substrat dans filtres de percolateur (BTF) dans les conditions ambiantes. Différents composés de soufre (sulfate, thiosulfate et soufre élémentaire) ont été utilisés comme accepteurs d'électrons. Lorsque le thiosulfate était utilisé comme accepteur d'électrons, sa dismutation aux sulfates et aux sulfures était la conversion dominante du soufre, mais aussi les taux les plus élevés d'AOM-SR étaient enregistrés dans ce LB. Par conséquent, l'AOM peut être directement couplé à la réduction du thiosulfate, ou à la réduction du sulfate produit par la dismutation du thiosulfate. De plus, l'utilisation de thiosulfate a conduit à l'enrichissement des SRB. Différemment, l'enrichissement le plus élevé d'ANME a été obtenu lorsque seul le sulfate était utilisé comme accepteur d'électrons.

Dans le BTF avec sulfate comme accepteur d'électrons, ANME et SRB ont été enrichis à partir de sédiments marins et les flux de carbone dans les microorganismes enrichis ont été étudiés par hybridation *in situ* en fluorescence et spectrométrie de masse à ionisation secondaire à l'échelle nanométrique (FISH-NanoSIMS). Les résultats préliminaires ont montré l'absorption de méthane par un groupe spécifique de SRB.

ANME et SRB adaptés aux sédiments profonds ont été enrichis en BTF à pression et température ambiantes. Le LB est un bioréacteur approprié pour l'enrichissement de micro-organismes à croissance lente. De plus, le thiosulfate et le sulfate peuvent être utilisés stratégiquement comme accepteurs d'électrons pour activer le sédiment et enrichir plus rapidement la population de SRB et ANME pour des applications futures.

# CHAPTER 1

## General Introduction and Thesis Outline

## 1.1 General introduction and problem statement

Methane ($CH_4$) is the most abundant hydrocarbon in the atmosphere and an important greenhouse gas, which has so far contributed to an estimated 20% of post-industrial global warming. The concentration of $CH_4$ in the atmosphere has been increasing at alarming rates and reducing $CH_4$ emission is thus important (Conrad, 2009; Kirschke et al., 2013). Ocean sediments produce large quantities of $CH_4$ by the methanogenic degradation of organic matter buried under the anoxic sea floor and an annual methanogenesis rate of 85-300 Tg $CH_4$ year$^{-1}$ has been estimated (Reeburgh, 2007). However, the ocean is also a major sink of $CH_4$, since most of the $CH_4$ produced is mainly oxidized before it reaches the hydrosphere and the atmosphere, of which more than 90% is consumed by anaerobic oxidation of $CH_4$ (AOM) (Hinrichs & Boetius, 2002; Reeburgh, 2007). AOM is restricted to anaerobic habitats such as, deep ocean, lake sediments and peats, mainly correlated to the reduction of sulfate, which is present in large quantities in the water column (~29 mM in seawater). AOM covers a wide range of rates, ranging from a few pmol cm$^{-3}$ day$^{-1}$ in the subsurface sulfate methane transition zone (SMTZ) of deep marine margins, to a few µmol cm$^{-3}$ day$^{-1}$ in surface sediments above gas hydrates. Moreover, AOM is evaluated to consume between 5 and 20% of the net atmospheric $CH_4$ flux (20-100 × 10$^{12}$ g year$^{-1}$) (Valentine et al., 2000).

AOM coupled to the reduction of sulfate (AOM-SR) is a process mediated by a consortium of anaerobic methanotrophic archaea (ANME) and sulfate reducing bacteria (SRB). So far, three types of ANME have been identified for AOM depending on the use of sulfate as the terminal electron acceptor (ANME-1, ANME-2, ANME-3) (Boetius et al., 2000; Knittel et al., 2005). Sulfate reducing AOM is a well-established phenomenon amongst deep marine environments; nevertheless, till to date the mechanism has not been fully understood and the cooperative/synergistic interaction between ANME and SRB is still under debate. Milucka et al. (2012) stated that a syntrophic partner might not be needed for ANME-2, while recent studies have shown the interactions between the two partners by direct electron transfer (McGlynn et al., 2015; Wegener et al., 2016) and that they can be decoupled by using external electron acceptors (Scheller et al., 2016). Moreover, recent studies have proved that AOM coupled to nitrite (Ettwig et al., 2010) and nitrate (Haroon et al., 2013) reduction, but also to iron and manganese reduction (Beal et al., 2009; Raghoebarsing et al., 2006) can occur, which are more favorable electron acceptors than sulfate. Besides, other sulfur compounds could also be used as electron acceptors for AOM: thiosulfate and sulfite are more thermodynamically favorable than sulfate (Meulepas et al., 2009b) and elemental sulfur can presumably be used directly by ANME (Milucka et al. 2012).

Most of the previous literature reports focused on the investigation of the microbial community involved in the AOM-SR process to understand the mechanism and to isolate the archaea involved. In all those studies, the main purpose was to define and understand the natural and biochemical cycle of $CH_4$. AOM investigation has another research direction, i.e. the desulfurization of industrial wastewater by using $CH_4$ as the sole electron donor. Sulfate and other sulfur oxyanions, such as thiosulfate, sulfite or dithionite, are contaminants discharged in fresh water due to industrial activities such as food processing, fermentation, coal mining, tannery and paper processing. Biological desulfurization under anaerobic conditions is a well-known biological treatment, in which these sulfur oxyanions are anaerobically reduced to sulfide (Liamleam & Annachhatre, 2007; Sievert et al., 2007; Weijma et al., 2006). The produced sulfide can immobilize toxic metals and decrease their bioavailability. In the process of wastewater desulfurization, electron donors for sulfate reduction are essential. Electron donors such as ethanol, hydrogen, methanol, acetate, lactate and propionate (Liamleam & Annachhatre, 2007) are usually supplied, but these increase the operational and investment costs (Meulepas et al., 2010). Therefore, the use of easily accessible and low-priced electron donors such as $CH_4$ is appealing for field-scale applications (Gonzalez-Gil et al., 2011).

So far, only a few researchers have attempted to study the process of AOM in bioreactors (Meulepas et al., 2009a; Zhang et al., 2010), with the main purpose of using $CH_4$ as the sole electron donor for sulfate removal from wastewater. In these bioreactors, marine sediment was used as the inoculum and they succeeded to achieve considerable higher AOM rates (0.6 mmol $l^{-1}$ $day^{-1}$) than those found in natural environments (3 $\mu$mol $l^{-1}$ $day^{-1}$). However, these studies have shown some constraints for practical application: (i) the slow growing nature of ANME in bioreactors, with least doubling time of 1.5-7.0 months and (ii) the low solubility of methane at ambient pressure (1.3 mM in seawater at 15ºC). These problems can be minimized by different strategies; by the enrichment of microorganisms in a bioreactor with a high biomass retention capability, by using high-pressure reactors for high methane solubility, by using alternative electron acceptors (elemental sulfur, thiosulfate) and electron donors for methanotrophs (acetate, ethanol) or by using microbial mats obtained from marine sediments where an active AOM is observed. Gulf of Cadiz sediment from mud volcanoes and mounds (e. g. Alpha Mound) are well known habitats for ANME and SRB (Niemann et al., 2006; Templer et al., 2011) and can be used as inoculum for their enrichment. Moreover, recent studies showed that marine Lake Grevelingen in the Netherlands with a water depth of just 45 m has a methane rich sediment (Egger et al., 2016), which hosts both ANME and SRB (Bhattarai et al., 2017) and therefore can be potentially used as inoculum for AOM-SR bioreactor.

The origin of the marine sediment, the methane availability, the substrates available and the type of bioreactor were considered in the approaches proposed in this research with the objective of controlling this natural phenomenon in a bioreactor in order to get more insight on the mechanism of AOM-SR and develop suitable strategy for the enrichment of the AOM community at ambient conditions for future applications.

## 1.2 Objectives and scope of the study

This research investigated new approaches to control AOM-SR and enrich ANME and SRB with the final purpose of designing a suitable bioreactor for AOM-SR at ambient pressure and temperature. The specific objectives of this research are:

1. Assess the factors controlling the distribution of ANME and the available tools for their enrichment *in vitro*

2. Investigate the effect of different sulfur compounds as electron acceptors and alternative electron donors on the AOM-SR activity by sediment from the marine Lake Grevelingen

3. Investigate the effect of different pressure gradients on AOM-SR activity and AOM community by sediment from the marine Lake Grevelingen

4. Investigate the use of thiosulfate as electron acceptor for AOM

5. Evaluate a biotrickling filter at ambient conditions for the enrichment of the microbial community mediating AOM and the reduction of different sulfur compounds as electron acceptors

6. Present a new process mediated strategy to investigate the mechanism of AOM-SR for future industrial applications

## 1.3 Thesis outline

This PhD thesis is divided into seven chapters. The first chapter (Chapter 1) provides a brief overview of this research and the thesis as depicted in Figure 1.1.

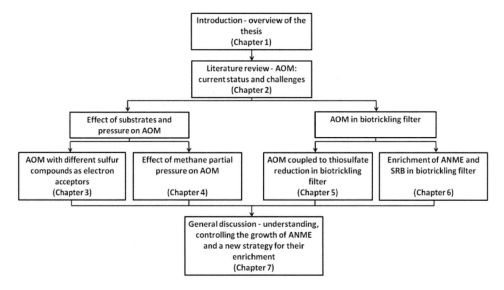

**Figure 1.1** Overview of the structure of this PhD thesis

Chapter 2 describes the current knowledge about AOM and the microorganisms involved, recent findings, the development of new study tools and their constraints. The recent findings about the cooperative interaction between ANME and SRB are summarized and discussed. Moreover, the distribution of ANME and the environmental factors responsible for this distribution are discussed in this chapter.

Chapter 3 discusses the effect of different substrates on AOM-SR by marine Lake Grevelingen sediment. The activity assays were performed in batches using different sulfur compounds as electron acceptors and different carbon sources as electron donors.

In Chapter 4, the microbes adapted to the shallow marine Lake Grevelingen sediment were subjected to different methane partial pressures. The effect of methane bioavailability and pressure on the ANME and SRB community was evaluated.

Chapter 5 focuses on the AOM coupled to thiosulfate reduction. Alpha Mound (Gulf of Cadiz) sediment was used as inoculum in a biotrickling filter. The reactions of the sulfur compounds involved were studied and the enriched microorganisms were visualized and quantified.

Chapter 6 gives the synthesis of the effect of different sulfur compounds as electron acceptors on the AOM-SR rates and on the ANME and SRB community adapted to the Alpha Mound sediment. A biotrickling filter was used for the enrichment of ANME and SRB at ambient temperature and pressure and the operational advantages of using this bioreactor for AOM-SR was discussed in this chapter. Moreover, the AOM activities were established and the enriched microorganisms were visualized and identified.

Chapter 7 provides a general discussion and outlook based on the specific research objectives of this thesis and it suggests new strategies for the enrichment of the AOM community in bioreactors. This chapter presents an overview on the practical application of this research and future research directions for investigating the AOM-SR mechanism has been presented in this chapter.

## 1.4 References

Beal, E.J., House, C.H., Orphan, V.J. 2009. Manganese- and iron-dependent marine methane oxidation. *Science*, **325**(5937), 184-187.

Bhattarai, S., Cassarini, C., Gonzalez-Gil, G., Egger, M., Slomp, C.P., Zhang, Y., Esposito, G., Lens, P.N.L. 2017. Anaerobic methane-oxidizing microbial community in a coastal marine sediment: anaerobic methanotrophy dominated by ANME-3. *Microb. Ecol.* **74**(3), 608-622.

Boetius, A., Ravenschlag, K., Schubert, C.J., Rickert, D., Widdel, F., Gieseke, A., Amann, R., Jørgensen, B.B., Witte, U., Pfannkuche, O. 2000. A marine microbial consortium apparently mediating anaerobic oxidation of methane. *Nature*, **407**(6804), 623-626.

Conrad, R. 2009. The global methane cycle: recent advances in understanding the microbial processes involved. *Environ. Microbiol. Rep.*, **1**(5), 285-292.

Egger, M., Lenstra, W., Jong, D., Meysman, F.J., Sapart, C.J., van der Veen, C., Röckmann, T., Gonzalez, S., Slomp, C.P. 2016. Rapid sediment accumulation results in high methane effluxes from coastal sediments. *PloS ONE*, **11**(8), e0161609.

Ettwig, K.F., Butler, M.K., Le Paslier, D., Pelletier, E., Mangenot, S., Kuypers, M.M.M., Schreiber, F., Dutilh, B.E., Zedelius, J., de Beer, D., Gloerich, J., Wessels, H.J.C.T., van Alen, T., Luesken, F., Wu, M.L., van de Pas-Schoonen, K.T., Op den Camp, H.J.M., Janssen-Megens, E.M., Francoijs, K.-J., Stunnenberg, H., Weissenbach, J.,

Jetten, M.S.M., Strous, M. 2010. Nitrite-driven anaerobic methane oxidation by oxygenic bacteria. *Nature*, **464**(7288), 543-548.

Gonzalez-Gil, G., Meulepas, R.J.W., Lens, P.N.L. 2011. Biotechnological aspects of the use of methane as electron donor for sulfate reduction. in: Murray, M.-Y. (Ed.), *Comprehensive Biotechnology*, Vol. 6 (2nd edition), Elsevier B.V. Amsterdam, the Netherlands, pp. 419-434.

Haroon, M.F., Hu, S., Shi, Y., Imelfort, M., Keller, J., Hugenholtz, P., Yuan, Z., Tyson, G.W. 2013. Anaerobic oxidation of methane coupled to nitrate reduction in a novel archaeal lineage. *Nature*, **500**(7468), 567-570.

Hinrichs, K.-U., Boetius, A. 2002. The anaerobic oxidation of methane: new insights in microbial ecology and biogeochemistry. in: Wefer, G., Billett, D., Hebbeln, D., Jørgensen, B.B., Schlüter, M., van Weering, T.E. (Eds), *Ocean Margin Systems*, Springer Berlin Heidelberg, Germany, pp. 457-477.

Kirschke, S., Bousquet, P., Ciais, P., Saunois, M., Canadell, J.G., Dlugokencky, E.J., Bergamaschi, P., Bergmann, D., Blake, D.R., Bruhwiler, L., Cameron-Smith, P., Castaldi, S., Chevallier, F., Feng, L., Fraser, A., Heimann, M., Hodson, E.L., Houweling, S., Josse, B., Fraser, P.J., Krummel, P.B., Lamarque, J.-F., Langenfelds, R.L., Le Quere, C., Naik, V., O'Doherty, S., Palmer, P.I., Pison, I., Plummer, D., Poulter, B., Prinn, R.G., Rigby, M., Ringeval, B., Santini, M., Schmidt, M., Shindell, D.T., Simpson, I.J., Spahni, R., Steele, L.P., Strode, S.A., Sudo, K., Szopa, S., van der Werf, G.R., Voulgarakis, A., van Weele, M., Weiss, R.F., Williams, J.E., Zeng, G. 2013. Three decades of global methane sources and sinks. *Nat. Geosci.*, **6**(10), 813-823.

Knittel, K., Lösekann, T., Boetius, A., Kort, R., Amann, R. 2005. Diversity and distribution of methanotrophic archaea at cold seeps. *Appl. Environ. Microbiol.*, **71**(1), 467-479.

Liamleam, W., Annachhatre, A.P. 2007. Electron donors for biological sulfate reduction. *Biotechnol. Adv.*, **25**(5), 452-463.

McGlynn, S.E., Chadwick, G.L., Kempes, C.P., Orphan, V.J. 2015. Single cell activity reveals direct electron transfer in methanotrophic consortia. *Nature*, **526**(7574), 531-535.

Meulepas, R.J.W., Jagersma, C.G., Gieteling, J., Buisman, C.J.N., Stams, A.J.M., Lens, P.N.L. 2009a. Enrichment of anaerobic methanotrophs in sulfate-reducing membrane bioreactors. *Biotechnol. Bioeng.*, **104**(3), 458-470.

Meulepas, R.J.W., Jagersma, C.G., Khadem, A.F., Buisman, C.J.N., Stams, A.J.M., Lens, P.N.L. 2009b. Effect of environmental conditions on sulfate reduction with methane as electron donor by an Eckernförde Bay enrichment. *Environ. Sci. Technol.*, **43**(17), 6553-6559.

Meulepas, R.J.W., Stams, A.J.M., Lens, P.N.L. 2010. Biotechnological aspects of sulfate reduction with methane as electron donor. *Rev. Environ. Sci. Biotechnol.*, **9**(1), 59-78.

Milucka, J., Ferdelman, T.G., Polerecky, L., Franzke, D., Wegener, G., Schmid, M., Lieberwirth, I., Wagner, M., Widdel, F., Kuypers, M.M.M. 2012. Zero-valent sulphur is a key intermediate in marine methane oxidation. *Nature*, **491**(7425), 541-546.

Niemann, H., Duarte, J., Hensen, C., Omoregie, E., Magalhães, V.H., Elvert, M., Pinheiro, L.M., Kopf, A., Boetius, A. 2006. Microbial methane turnover at mud volcanoes of the Gulf of Cadiz. *Geochim. Cosmochim. Ac.*, **70**(21), 5336-5355.

Raghoebarsing, A.A., Pol, A., van de Pas-Schoonen, K.T., Smolders, A.J.P., Ettwig, K.F., Rijpstra, W.I.C., Schouten, S., Damsté, J.S.S., Op den Camp, H.J.M., Jetten, M.S.M., Strous, M. 2006. A microbial consortium couples anaerobic methane oxidation to denitrification. *Nature*, **440**(7086), 918-921.

Reeburgh, W.S. 2007. Oceanic methane biogeochemistry. *Chem. Rev.*, **107**(2), 486-513.

Scheller, S., Yu, H., Chadwick, G.L., McGlynn, S.E., Orphan, V.J. 2016. Artificial electron acceptors decouple archaeal methane oxidation from sulfate reduction. *Science*, **351**(6274), 703-707.

Sievert, S.M., Kiene, R.P., Schulz-Vogt, H.N. 2007. The sulfur cycle. *Oceanography*, **20**(2), 117-123.

Templer, S.P., Wehrmann, L.M., Zhang, Y., Vasconcelos, C., McKenzie, J.A. 2011. Microbial community composition and biogeochemical processes in cold-water coral carbonate mounds in the Gulf of Cadiz, on the Moroccan margin. *Mar. Geol.*, **282**(1-2), 138-148.

Valentine, D.L., Reeburgh, W.S., Hall, R. 2000. New perspectives on anaerobic methane oxidation. *Environ. Microbiol.*, **2**(5), 477-484.

Wegener, G., Krukenberg, V., Ruff, S.E., Kellermann, M.Y., Knittel, K. 2016. Metabolic capabilities of microorganisms involved in and associated with the anaerobic oxidation of methane. *Front. Microbiol.*, **7**(46), 1-16.

Weijma, J., Veeken, A., Dijkman, H., Huisman, J., Lens, P.N.L. 2006. Heavy metal removal with biogenic sulphide: advancing to full-scale. in: Cervantes, F., Pavlostathis, S., van Haandel, A. (Eds.), *Advanced biological treatment processes for industrial wastewaters, principles and applications*, IWA publishing. London, pp. 321-333.

Zhang, Y., Henriet, J.-P., Bursens, J., Boon, N. 2010. Stimulation of *in vitro* anaerobic oxidation of methane rate in a continuous high-pressure bioreactor. *Bioresour. Technol.*, **101**(9), 3132-3138.

# CHAPTER 2

# Physiology and Distribution of Anaerobic Oxidation of Methane by Archaeal Methanotrophs

## Abstract

Methane is oxidized in marine anaerobic environments, where sulfate rich sea water meets biogenic or thermogenic methane. In those niches, few phylogenetically distinct microbial types, i.e. anaerobic methanotrophs (ANME), are able to grow through anaerobic oxidation of methane (AOM). Due to the relevance of methane in the global carbon cycle, ANME draw the attention of a broad scientific community since five decades. This review presents and discusses the microbiology and physiology of ANME up to the recent discoveries, revealing novel physiological types of anaerobic methane oxidizers which challenge the view of obligate syntrophy for AOM. The drivers shaping the distribution of ANME in different marine habitats, from cold seep sediments to hydrothermal vents, are overviewed. Multivariate analyses of the abundance of ANME in various habitats identify a distribution of distinct ANME types driven by the mode of methane transport. Intriguingly, ANME have not yet been cultivated in pure culture, despite of intense attempts. Further, advances in understanding this microbial process are hampered by insufficient amounts of enriched cultures. This review discusses the advantages, limitations and potential improvements for ANME cultivation systems and AOM study approaches.

## 2.1 Introduction

Methane ($CH_4$) is the most abundant and completely reduced form of hydrocarbon. It is the most stable hydrocarbon, which demands $+439$ kJ mol$^{-1}$ energy to dissociate the hydrocarbon bond (Thauer & Shima, 2008). $CH_4$ is a widely used energy source, but it is also the second largest contributor to human induced global warming, after $CO_2$. $CH_4$ concentrations in the atmosphere have increased from about 0.7 to 1.8 ppmv (i.e. an increase of 150%) in last 200 years, and experts estimate that this increase is responsible for approximately 20% of the Earth's warming since pre-industrial times (Kirschke et al., 2013). On a per mol basis and over a 100-year horizon, the global warming potential of $CH_4$ is about 25 times more than that of $CO_2$ (IPCC, 2007). Therefore, large scientific efforts are being made to resolve detailed maps of $CH_4$ sources and sinks, and how these are affected by the increased levels of this gas in the atmosphere (Kirschke et al., 2013).

The global $CH_4$ cycle is largely driven by microbial processes of $CH_4$ production (i.e. methanogenesis) and $CH_4$ oxidation (i.e. methanotrophy). $CH_4$ is microbially produced by the anaerobic degradation of organic compounds or through $CO_2$ bioreduction (Nazaries et al., 2013). These $CH_4$ production processes occur in diverse anoxic subsurface environments like rice paddies, wetlands, landfills, contaminated aquifers as well as freshwater and ocean sediments (Reeburgh, 2007). $CH_4$ can also be formed physio-chemically at specific temperatures of about 150°C to 220°C (thermogenesis). It is estimated that more than half of the $CH_4$ produced globally is oxidized microbially to $CO_2$ before it reaches the atmosphere (Reeburgh, 2007). Both aerobic and anaerobic methanotrophy are the responsible processes. The first involves the oxidation of $CH_4$ to methanol in the presence of molecular oxygen (and subsequently to $CO_2$) by methanotrophic bacteria (Chistoserdova et al., 2005; Hanson & Hanson, 1996), whereas the second includes the oxidation of $CH_4$ to $CO_2$ in the absence of oxygen by a clade of archaea, called anaerobic methanotrophs (ANME) and the process is known as the anaerobic oxidation of $CH_4$ (AOM).

Large $CH_4$ reservoirs on Earth, from 450 to 10,000 Gt carbon (Gt C) (Archer et al., 2009; Wallmann et al., 2012) are found as $CH_4$ hydrates beneath marine sediments, mostly formed by biogenic processes (Pinero et al., 2013). $CH_4$ hydrates, or $CH_4$ clathrates, are crystalline solids, consisting of large amounts of $CH_4$ trapped by interlocking water molecules (ice). They are stable at high pressure (> 60 bar) and low temperature (< 4°C) (Boetius & Wenzhöfer, 2013; Buffett & Archer, 2004), and are typically found along continental margins at depths of 600 to 3000 m below sea level (Archer et al., 2009; Boetius & Wenzhöfer, 2013; Reeburgh, 2007). By gravitational and tectonic forces, $CH_4$ stored in hydrate seeps into the ocean sediment under the form of mud volcanoes, gas chimneys, hydrate mounds and pock marks (Boetius & Wenzhöfer, 2013). These $CH_4$ seepage manifestations are environments where AOM has been documented (Table 2.1) (e.g. Black Sea carbonate chimney (Treude et al., 2007), Gulf of Cadiz mud volcanoes (Niemann et al., 2006a), Gulf of Mexico gas hydrates (Joye et al., 2004). Besides, AOM also occurs in the sulfate-$CH_4$ transition zones (SMTZ) of sediments. The SMTZ are quiescent sediment environments, where the upwards diffusing (thermogenic and biogenic) $CH_4$ is oxidized when it meets sulfate ($SO_4^{2-}$), which is transported downwards from the overlaying seawater (Figure 2.1). Considering that $SO_4^{2-}$ is abundant in seawater and that oxygen in sea bed sediments is almost absent, AOM coupled to the reduction of $SO_4^{2-}$ is likely the dominant biological sink of $CH_4$ in these environments.

It is estimated that $CH_4$ seeps, which are generally laying above $CH_4$ hydrates (Suess, 2014), annually emit 0.01 to 0.05 Gt C, contributing to 1 to 5 % of the global $CH_4$ emissions to the atmosphere (Boetius & Wenzhöfer, 2013). These emissions would be higher if $CH_4$ was not scavenged by aerobic or anaerobic oxidation of $CH_4$. While aerobic $CH_4$ oxidation is dominant in shallow oxic seawaters (Tavormina et al., 2010), AOM is found in the anoxic zones of the sea floor (Knittel & Boetius, 2009; Reeburgh, 2007; Wankel et al., 2010). Due to limited data, it has not been possible to determine the exact global values of $CH_4$ consumption by AOM. But, the AOM in the SMTZ and $CH_4$ seep environments has been tentatively estimated at 0.05 Gt C and 0.01 Gt C per year, respectively (Boetius & Wenzhöfer, 2013).

Besides the biogeochemical implications of AOM, this microbial process can have biotechnological applications for the treatment of waste streams rich in $SO_4^{2-}$ or nitrate/nitrite but low in electron donor. Recently, a few studies have highlighted the prospective of AOM and ANME in environmental biotechnology, where $CH_4$ is used as the sole electron donor to achieve $SO_4^{2-}$ reduction (SR) in bioreactors (Gonzalez-Gil et al., 2011; Meulepas et al., 2009a; Meulepas et al., 2010c). Biological SR is a well-known technique to remove sulfur and metals from wastewaters, metals can be recovered by metal sulfide precipitation (Lens et al., 2002). Many industrial wastewaters are deficient in dissolved organic carbon. Hence, supplementation of external carbon sources and electron donors is essential for microbial SR. Frequently used electron donors for $SO_4^{2-}$ reducing treatment plants are hydrogen/ $CO_2$ and ethanol (Widdel & Hansen, 1992), which are costly and can be replaced by low-priced electron donors (Gonzalez-Gil et al., 2011). It is estimated that the overall treatment costs would be reduced by a factor of

2 to 4 if $CH_4$ from natural gas or biogas would be used in $SO_4^{2-}$ reducing bioreactors as an electron donor instead of hydrogen or ethanol (Meulepas et al., 2010c). The major limitation identified for the biotechnological application of AOM is the extremely low growth rates of the ANME, currently with doubling times as high as 2-7 months (Meulepas et al., 2010c).

A recent innovative idea is the use of key AOM enzymes for the biotechnological conversion of $CH_4$ to liquid fuels at high carbon conservation efficiencies (Haynes & Gonzalez, 2014). $CH_4$ could be transformed into butanol efficiently, if enzymes responsible for AOM activate $CH_4$ and assist in C-C bond formation (Haynes & Gonzalez, 2014). This concept is of interest because logistics and infrastructure for handling liquid fuels are more cost effective than those for utilizing compressed natural gas. A detailed elucidation of the ANME metabolism is a prerequisite to the development of such biotechnological applications of AOM.

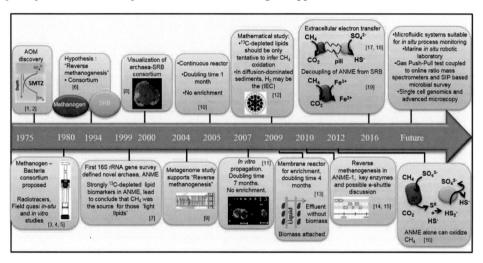

**Figure 2.1** Timeline of relevant research and discoveries on the AOM-SR. The major milestones achieved are depicted in their respective year along with some future possibility in the AOM studies. Note for references: 1-(Reeburgh, 1976), 2-(Martens & Berner, 1974), 3-(Zehnder & Brock, 1980),4-(Iversen & Jørgensen, 1985), 5-(Reeburgh, 1980), 6-(Hoehler et al., 1994), 7-(Hinrichs et al., 1999), 8-(Boetius et al., 2000), 9-(Hallam et al., 2004), 10-(Girguis et al., 2005), 11-(Nauhaus et al., 2007), 12-(Alperin & Hoehler, 2009), 13-(Meulepas et al., 2009a), 14-(Scheller et al., 2010), 15-(Meyerdierks et al., 2010), 16-(Milucka et al., 2012) , 17-(McGlynn et al., 2015), 18-(Wegener et al., 2016), 19-(Scheller et al., 2016)

## 2.2 Microbiology of AOM

### 2.2.1 Discovery of AOM

AOM coupled to $SO_4^{2-}$ reduction (AOM-SR) takes place where $SO_4^{2-}$ meets either biogenic or thermogenic $CH_4$. This unique microbiological phenomenon, AOM, was recognized since four decades as a key to close the balance of oceanic carbon (Martens & Berner, 1974; Reeburgh, 1976). Since then, various key discoveries have elucidated the AOM process to some extent, but its exact biochemical mechanism is still unclear (Figure 2.1). The AOM was first deduced from $CH_4$ and $SO_4^{2-}$ profile measurements in marine sediments (Iversen & Jørgensen, 1985; Reeburgh, 1980; Zehnder & Brock, 1980). Occurrence of AOM yields typical concave-up $CH_4$ profiles in sediment columns with high $CH_4$ concentrations in the deep sediment layers and very low $CH_4$ concentrations at the sediment water interface (Figure 2.1).

Quasi *in situ* and *in vitro* studies using radiotracers confirmed AOM as a biological process (Iversen & Jørgensen, 1985; Reeburgh, 1980; Zehnder & Brock, 1980). Additional *in vitro* studies suggested that the AOM process was performed by a unique microbial community (Boetius et al., 2000; Hoehler et al., 1994): the anaerobic methanotrophs (ANME), mostly in association with $SO_4^{2-}$ reducing bacteria (SRB) (Figure 2.2). The identification of the microorganisms involved in AOM is crucial to explain how $CH_4$ can be efficiently oxidized with such a low energy yield. By fluorescence *in situ* hybridization (FISH) based visualizations with specifically designed probes, the *in situ* occurrence of such archaea-bacteria associations was recorded, showing that the ANME-groups are widely distributed throughout marine sediments (Boetius et al., 2000; Hinrichs et al., 1999; Knittel et al., 2005; Orphan et al., 2002; Schreiber et al., 2010). The physico-chemical drivers shaping the global distribution of ANME consortia are not fully resolved to date (section 2.4). Instead, AOM activity tests and *in vitro* studies allowed the estimation of their doubling time in the order of 2-7 months, realizing the extremely slow growth of ANME on $CH_4$ (Nauhaus et al., 2007).

## 2.2.2 ANME phylogeny

Based on the phylogenetic analysis of 16S rRNA genes (Figure 2.3A), ANME have been grouped into three distinct clades, i.e. ANME-1, ANME-2 and ANME-3 (Boetius et al., 2000; Hinrichs et al., 1999; Knittel et al., 2005; Niemann et al., 2006b). All ANME are phylogenetically related to various groups of methanogenic Archaea (Figure 2.3). ANME-2 and ANME-3 are clustered within the order *Methanosarcinales*, while ANME-1 belongs to a new order which is distantly related to the orders *Methanosarcinales* and *Methanomicrobiales* (Figure 2.3). Specifically, ANME-3 is closely related to the genus *Methanococcoides*. FISH analysis showed that microorganisms belonging to the ANME-2 and ANME-3 groups are cocci-shaped, similar to *Methanosarcina* and *Methanococcus* methanogens (Figures 2.2B and 2.2D). On the contrary to ANME-2 and ANME-3, ANME-1 mostly exhibits a rod-shape morphology (Figure 2.2A).

Upon phylogenetic analysis based on 16S rRNA (Figure 2.3A) and *mcr*A (Figure 2.3B) genes, the three major groups of ANME were identified. ANME-1 is further subgrouped into ANME-1a and ANME-1b. ANME-2 is divided into four subgroups, i.e. ANME-2a, ANME-2b, ANME-2c and ANME-2d, whereas, so far no subgroups of ANME-3 have been defined (Figure 2.3). The *mcr*A genes phylogeny of the various archaeal orders closely parallels that of the 16S rRNA genes (Figure 2.3).

Besides their close phylogenetic relationships, ANME exhibit other similarities with methanogenic archaea. For example, sequenced genomes of ANME-1 and ANME-2 from environmental samples indicate that, except for the $N^5$, $N^{10}$-methylene-tetrahydromethanopterin ($H_4$MPT) reductase in the ANME-1 metagenome (Meyerdierks et al.,

2010), these ANME contain homologous genes for the enzymes involved in all the seven steps of methanogenesis from $CO_2$ (Haroon et al., 2013; Meyerdierks et al., 2010; Wang et al., 2014). Furthermore, with the exception of coenzyme M-S-S-coenzyme B heterodisulfide reductase, all those enzymes catalyzing the $CH_4$ formation were confirmed to catalyze reversible reactions (Rudolf, 2011; Scheller et al., 2010). Thus, it is hypothesized that ANME oxidize $CH_4$ via methanogenic enzymatic machinery functioning in reverse, i.e., reversal of $CO_2$ reduction to $CH_4$ (Hallam et al., 2004; Meyerdierks et al., 2010).

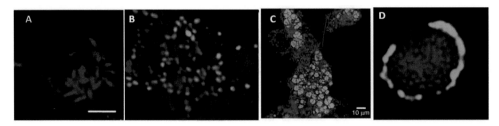

**Figure 2.2** Fluorescence *in situ* hybridization images from different ANME types. A) Single ANME-1 in elongated rectangular shape (red color) inhabiting as mono specific clade in the Guaymas Basin hydrothermal vent (Holler et al., 2011a), B) Aggregate of cocci shaped ANME-2 (red color) and *DSS* (green color), enrichment sample after 8 years from the Isis mud volcano in the Mediterranean Sea. The image was taken from the web: http://www.mpg.de/6619070/marine-CH4-oxidation, C) Aggregate of large densely clustered ANME-2d (green) and other bacteria (blue color) obtained from a bioreactor enrichment (Haroon et al., 2013) and D) Aggregate of cocci shaped ANME-3 (red color) and *DBB* (green color) inhabiting Haakon Mosby mud volcano (Niemann et al., 2006b).

### 2.2.3   AOM coupled to sulfate reduction

The ocean is one of the main reservoirs of sulfur, where it mainly occurs as dissolved $SO_4^{2-}$ in seawater or as mineral in the form of pyrite ($FeS_2$) and gypsum ($CaSO_4$) in sediments (Sievert et al., 2007). Sulfur exists in different oxidation states, with sulfide ($S^{2-}$), elemental sulfur ($S^0$) and $SO_4^{2-}$ as the most abundant and stable species in nature. With an amount of 29 mM, $SO_4^{2-}$ is the most dominant anion in ocean water, next to chloride. The sedimentary sulfur cycle involves two main microbial processes: (i) bacterial dissimilatory reduction of $SO_4^{2-}$ to hydrogen sulfide, which can subsequently precipitate with metal ions (mainly iron), and (ii) assimilatory reduction of $SO_4^{2-}$ to form organic sulfur compounds incorporated in microbial biomass (Jørgensen & Kasten, 2006). Dissimilatory SR by SRB occurs in anoxic marine sediments or in freshwater environments, where SRB use several electron donors, such as hydrogen, various organic compounds (e.g. ethanol, formate, lactate, pyruvate, fatty acids, methanol, and methanethiol) as well as $CH_4$ (Muyzer & Stams, 2008).

AOM was considered impossible in the past, due to the non polar C-H bond of $CH_4$ (Thauer & Shima, 2008). From a thermodynamic point of view, AOM-SR yields minimal energy: only 16.6 kJ $mol^{-1}$ of energy is released during AOM-SR (Eq. 2.1 in Figure 2.4). In comparison, more energy is released by the hydrolysis of one ATP (31.8 kJ $mol^{-1}$). Other electron acceptors in the anaerobic environment, such as nitrate, iron and manganese provide higher energy yields than $SO_4^{2-}$, as deduced by the $\Delta G^{0'}$ of the different redox reactions (Figure 2.4). However, their combined concentration at the marine sediment-water interface is far lower than the $SO_4^{2-}$ concentration (D'Hondt et al., 2002). Therefore, AOM-SR usually dominates in marine sediments.

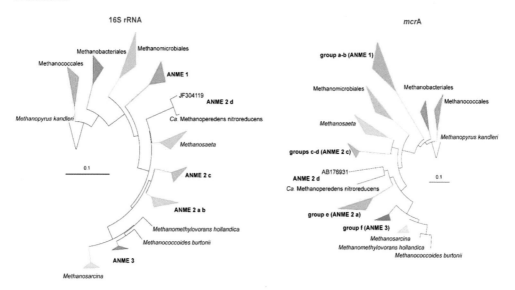

**Figure 2.3** Phylogenetic affiliation of anaerobic methanotrophs (ANME) based on the 16S rRNA and *mcr*A genes. The 16S rRNA and *mcr*A sequences, retrieved from NCBI databases (Pruesse et al., 2012), were respectively aligned with the SINA aligner and the Clustal method as previously described (Hallam et al., 2003). Both trees were inferred using the Neighbor-Joining method (Saitou & Nei, 1987). Bars refer to 10% estimated distance.

AOM-SR was suggested to be a cooperative metabolic process of the AOM coupled to dissimilatory $SO_4^{2-}$ reduction, thereby gaining energy by a syntrophic consortium of ANME and SRB (Boetius et al., 2000; Hoehler et al., 1994) (Eq. 2.1 in Figure 2.4). Especially the *Desulfosarcina / Desulfococcus* (*DSS*) and *Desulfobulbaceae* (*DBB*) clades of SRB are common associates of ANME for SR. However, the three ANME phylotypes have been visualized without any attached SRB in different marine environments as well (Losekann et al., 2007; Maignien et al., 2013; Treude et al., 2005a; Wankel et al., 2012a), suggesting that AOM-SR can potentially be performed independently by the ANME themselves (Eq. 2.1 Figure 2.4, performed solely by ANME). Theoretically, slightly more energy can be released (18 kJ mol⁻¹) if $SO_4^{2-}$ is reduced to disulfide instead of sulfide (Eq. 2.2 in Figure 2.4) (Milucka et al., 2012).

### 2.2.4 AOM coupled to different sulfur compounds as electron acceptors

Microorganisms that mediate AOM-SR can also use $S^0$ or thiosulfate ($S_2O_3^{2-}$) as terminal electron acceptor for AOM (Figure 2.4, Eq. 2.7 and Eq. 2.8). The reduction of one mole of $S_2O_3^{2-}$ to one mole sulfide requires fewer electrons (4 electrons) than the reduction of $SO_4^{2-}$ to sulfide (8 electrons). The reduction of $S_2O_3^{2-}$ coupled to $CH_4$ oxidation is also more energetically favorable (Eq. 2.8 in Figure 2.4) than AOM-SR (Eq. 2.1 in Figure 2.4). However, researches investigating AOM coupled to $S_2O_3^{2-}$ reduction (Suarez-Zuluaga et al., 2015; Suarez-Zuluaga et al., 2014), showed that $S_2O_3^{2-}$ disproportionation prevailed over its reduction, even if it is theoretically less thermodynamically favorable ($\Delta G^{0'}$ = -22 kJ mol$^{-1}$). The presence of known SRB able to metabolize inorganic sulfur compounds by disproportionation, such as *Desulfocapsa* and *Desulfovibrio* (Finster, 2008), in the studied sediment (Suarez-Zuluaga et al., 2015) might favor $S_2O_3^{2-}$ over its reduction with $CH_4$ as sole electron donor. However, the *DSS* and *DBB* commonly associated to ANME were never proved to metabolize $S_2O_3^{2-}$ disproportion, even though *DSS* were once described as putative disulfide disproportionating bacteria (Milucka et al., 2012).

Differently, the theoretical Gibbs free energy for AOM coupled to $S^0$ is positive (+24 kJ mol$^{-1}$, Eq. 2.7 in Figure 2.4) however, *in vitro* tests showed that this reaction may well proceed and the calculated free energy of reaction at *in situ* conditions is negative (-84.1 kJ mol$^{-1}$) (Milucka et al., 2012). Contrarily than for $S_2O_3^{2-}$, $S^0$ disproportionation requires energy (+41 kJ mol$^{-1}$) unless an oxidant, as Fe (III), renders the reaction more energetically favorable (Finster, 2008) or in alkaline environments, such as soda lakes (Poser et al., 2013). Therefore, other reactions and mechanisms might be taken into consideration when investigating AOM coupled to the reduction of other sulfur compounds asuch as $S_2O_3^{2-}$ and $S^0$.

### 2.2.5   AOM coupled to nitrite and nitrate reduction

Methanotrophs that utilize nitrite (Ettwig et al., 2010) or nitrate (Haroon et al., 2013) have been identified in anaerobic fresh water sediments. Thermodynamically, the AOM coupled to nitrite and nitrate yields more energy than AOM-SR, with a $\Delta G^0$ of -990 kJ mol$^{-1}$ and -785 kJ mol$^{-1}$, respectively (Eq. 2.5 and Eq. 2.6 in Figure 2.4). Two specific groups of microbes are involved in the process of AOM coupled with nitrate and nitrite reduction: "*Candidatus Methanoperedens nitroreducens*" (archaea) and "*Candidatus Methylomirabilis oxyfera*" (bacteria), respectively.

AOM coupled to denitrification was first hypothesized to occur in a similar syntrophic manner as AOM coupled to SR (Raghoebarsing et al., 2006). However, Ettwig *et al.* (Ettwig et al., 2010) showed that $CH_4$ oxidation coupled to nitrite reduction occurs in the absence of archaea. The bacterium "*Candidatus Methylomirabilis oxyfera*" couples AOM to denitrification, with nitrite being reduced to nitric oxide which is then converted to nitrogen ($N_2$) and oxygen ($O_2$). The thus generated intracellular byproduct oxygen is subsequently used to oxidize $CH_4$ to $CO_2$ (Ettwig et al., 2010). Moreover, recent studies reveal that a distinct ANME, affiliated to the ANME-2d subgroup and named "*Candidatus Methanoperedens nitroreducens*" (Figures 2.2C and 2.3), can carry out AOM using nitrate as the terminal electron acceptor through reversed methanogenesis (Haroon et al., 2013). In the presence of ammonium, the nitrite released by this ANME-2d is then reduced to $N_2$ by the anaerobic ammonium-oxidizing bacterium (anammox) "*Candidatus Kuenenia* spp."; while in the absence of ammonium, nitrate is reduced to $N_2$ by "*Candidatus Methylomirabilis oxyfera*". Therefore, different co-cultures are dominated in a biological system depending on the availability of the nitrogen species (nitrate, nitrite or ammonium) (Haroon et al., 2013).

### 2.2.6 Other electron acceptors for AOM

Besides $SO_4^{2-}$ and nitrate, iron and manganese are other electron acceptors studied for AOM. In marine sediments, AOM was found to be coupled to the reduction of manganese or iron (Beal et al., 2009; Ettwig et al., 2016), but whether manganese and iron are directly used for the process or not, is yet to be elucidated. An *in vitro* study from Beal *et al.* (2009) showed that oxide minerals of manganese, birnessite (simplified as $MnO_2$ in Eq. 2.4 of Figure 2.4) and iron, ferrihydrite (simplified as $Fe(OH)_3$ in Eq. 2.3 of Figure 2.4), can be used as electron acceptors for AOM. The rates of AOM coupled to $MnO_2$ or $Fe(OH)_3$ reduction are lower than AOM-SR, but the energy yields ($\Delta G^{0'}$ of -774 kJ mol$^{-1}$ and -556 kJ mol$^{-1}$ respectively, Eq. 2.3 and Eq. 2.4 in Figure 2.4) are higher. Thus, the potential energy gain of Mn- and Fe-dependent AOM is, respectively, 10 and 2 times higher than that of AOM-SR, inspiring researchers to further investigate these potential processes (Beal et al., 2009).

Several researchers have investigated on the identity of the bacteria present in putative Fe- and Mn- dependent AOM sites and hypothesized their involvement along with ANME (Beal et al., 2009; Wankel et al., 2012a). In parallel to AOM-SR, this process is also assumed to be mediated by two cooperative groups of microorganisms. The bacterial 16S rRNA phylotypes found in Fe- and Mn-dependent AOM sites are putative metal reducers, belonging to the phyla *Verrucomicrobia phylotypes* (Wankel et al., 2012a), *Bacteriodetes*, *Proteobacteria* and *Acidobacteria* and are mostly present in heavy-metal polluted sites and hydrothermal vent systems (Beal et al., 2009). The latter bacteria are mostly present. The ANME-1 clade was identified as the most abundant in metalliferous hydrothermal sediments and in Eel River Basin $CH_4$-seep sediment. However, the sole identification of specific bacteria and archaea in these marine sediments does not provide evidence for their metal reducing capacity.

Recent studies assumed the direct coupling of AOM to iron reduction. Wankel et al. (2012a) investigated AOM in hydrothermal sediments from the Middle Valley vent field, where AOM occurred in the absence of SR and SRB. Fe-dependent AOM was hypothesized as the process in these sediments, due to the abundance of Fe (III)-bearing minerals, specifically green rust and a mixed ferrous-ferric hydroxide. A higher AOM rate than with SR was observed in *in vitro* incubations with Mn and Fe based electron acceptors like birnessite and ferrihydrite (Segarra et al., 2013). Moreover, Scheller et al. (2016) showed that marine samples containing ANME-2 could couple the reduction of chelated oxidized iron and recently Ettwig et al. (2016) demonstrates that iron and manganese dependent $CH_4$ oxidation occurred in a freshwater enrichment culture of *"Candidatus Methanoperedens nitroreducens"*.

There is also a hypothesis on possible indirect coupling of AOM with metal reduction (Beal et al., 2009). Namely, sulfide, present in the sediment, is oxidized to elemental sulfur and disulfide in the presence of metal oxides. The produced sulfur compounds can be disproportionated by bacteria producing transient $SO_4^{2-}$, which can be used to oxidize $CH_4$.

These sulfur transformations are referred to as cryptic sulfur cycling (Aller & Rude, 1988; Canfield et al., 1993) and its extent can increase if the sediment is rich in microorganisms able to metabolize elemental sulfur and disulfide (Straub & Schink, 2004; Wan et al., 2014). A recent study with the Bothnian Sea sediment speculated two separate anaerobic regions where AOM occurs: AOM-SR (in the upper anaerobic layer) and Fe-dependent AOM (in the lower anaerobic layer). It was hypothesized that the majority of AOM was coupled directly to iron reduction in the iron reducing region and only about 0.1% of AOM-SR was due to cryptic sulfur cycling (Egger et al., 2015). However, in marine and brackish sediments probably only a few percent of the $CH_4$ is oxidized by a Fe-dependent process.

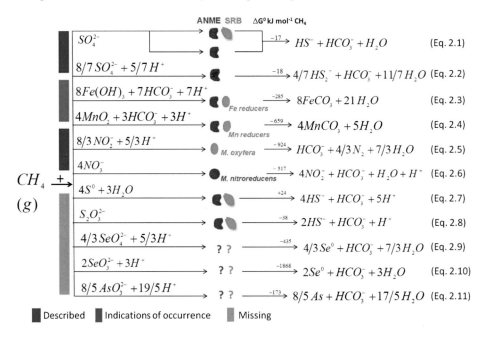

Figure 2.4 Described and possible AOM processes with different terminal acceptors. The AOM with $SO_4^{2-}$, nitrate and nitrite as electron acceptors is well described along with the microbes involved, which is indicated by green blocks, whereas the AOM with manganese and iron was shown but the microbes involved need to be characterized which is indicated by the blue block. Other possible electron acceptors are mentioned according to the thermodynamic calculation of the chemical reactions, which is indicated by the orange block.

Theoretically, based on thermodynamics, anaerobic $CH_4$ oxidizing microorganisms could utilize other electron acceptors including arsenic and selenium. It should be noted that the chemistry of selenium oxyanions is similar to that of sulfur oxyanions, since both belong to the same group in the periodic table, the so called chalcogens. Oxidized selenium species i.e. selenate or selenite, might thus also be used as electron acceptor for AOM (Eq. 2.9 and Eq. 2.10 in Figure 2.4).

## 2.3   Physiology of ANME

### 2.3.1   Carbon and nitrogen metabolism

The difficulty in obtaining enrichment cultures of ANME hampers getting insights into the physiological traits of these microorganisms. Nonetheless, *in situ* and *in vitro* activity tests using $^{13}$C- or $^{14}$C-labelled $CH_4$ unequivocally revealed that ANME oxidize $CH_4$ (Nauhaus et al., 2007). But the physiology of these microorganisms seems to be more intriguing. Recently, it was found that the carbon in ANME biomass is not totally derived from $CH_4$, i.e. ANME are not obligate heterotrophs. ANME-2 and their bacterial partners (Wegener et al., 2016) have been defined as autotrophic, whereas carbon within the biomass of ANME-1 is derived from $CO_2$ fixation (Kellermann et al., 2012; Treude et al., 2007). Furthermore, genetic studies showed that ANME-1 contains genes encoding the $CO_2$ fixation pathway characteristic for methanogens (Meyerdierks et al., 2010).

There is evidence that some ANME-1 and/or ANME-2 from the Black Sea and from the Gulf of Mexico $CH_4$ seeps can produce $CH_4$ (Orcutt et al., 2005; Treude et al., 2007) from $CO_2$ or from methanol (Bertram et al., 2013). This methanogenic capacity exhibited by these ANME seems in turn to mirror the $CH_4$ oxidation capacity displayed by pure cultures of methanogens (Harder, 1997; Zehnder & Brock, 1979) and by methanogens present in anaerobic sludge (Meulepas et al., 2010b), which can oxidize about 1 to 10% of the $CH_4$ they produce. However, the reported $CH_4$ oxidation capacity of cultured methanogens is so low that they are not considered to contribute to $CH_4$ oxidation in marine settings. On the contrary, the detection of important numbers of active ANME-1 cells in both the $CH_4$ oxidation and the $CH_4$ production zones of estuary sediments has led to the proposition that this ANME type is not an obligate $CH_4$ oxidizer, but rather a flexible type which can switch and function as methanogen as well (Lloyd et al., 2011).

Another intriguing physiological trait is the $N_2$ fixing capacity (i.e., diazotrophy) by ANME-2d. Using $^{15}N_2$ as nitrogen source, it was found that ANME-2d cells assimilated $^{15}N$ in batch incubations of marine mud volcano or $CH_4$ seep sediments (Dekas et al., 2014; Dekas et al., 2009). While fixing $N_2$, ANME maintained their $CH_4$ oxidation rate, but their growth rate was severely reduced. The energetic cost to fix nitrogen is one of the highest amongst all anabolic processes and requires about 16 ATP molecules, which translates into 800 kJ mol$^{-1}$ of nitrogen reduced. Therefore, considering the meager energy gain of AOM (about 30 or 18 kJ mol$^{-1}$ of $CH_4$ oxidized), it is consistent that the growth rate of ANME can be 20 times lower using $N_2$ than using ammonium ($NH_4^+$) as nitrogen source (Dekas et al., 2009). Yet, it is not resolved under which *in situ* conditions these microorganism would be diazotrophic. Also, whether other ANME types are diazotrophs has not yet been shown. Although the metagenome of ANME-1 reveals the presence of various candidate proteins having similarity to proteins known to be involved in $N_2$ fixation (Meyerdierks et al., 2010), this trait has not yet been tested experimentally.

### 2.3.2 Syntrophy and potential electron transfer modes between ANME and SRB

Several theories have been proposed to understand the mechanism between ANME archaea and their association with SRB, with the most common hypothesis of syntrophy between ANME and SRB (Figure 2.5A). The syntrophy between ANME and SRB is hypothesized on the basis of the tight co-occurrence of ANME and SRB in AOM active sites, as revealed by FISH images (Figures 2.2B and 2.2D) (Blumenberg et al., 2004; Boetius et al., 2000; Knittel et al., 2005), but also phylogenetic analysis showed the co-occurrence of SRB and ANME in samples from AOM sites (Alain et al., 2006; Losekann et al., 2007; Stadnitskaia et al., 2005). Obligate syntrophs usually share the substrate degradation process resulting in one partner converting the substrate into an intermediate, which is consumed by the syntrophic partner (Stams & Plugge, 2009). Unlike other known forms of syntrophy, the intermediate shared by ANME and SRB has not yet been identified. Isotopic signatures in archaeal and bacterial lipid biomarker based analysis strengthened this hypothesis, assuming transfer of an intermediate substrate between the two microorganisms (Boetius et al., 2000; Hinrichs & Boetius, 2003; Hinrichs et al., 2000).

Hydrogen and other methanogenic substrates, such as acetate, formate, methanol and methanethiol were hypothesized as the intermediates between ANME and SRB (Figure 2.5A) (Hoehler et al., 1994; Sørensen et al., 2001; Valentine et al., 2000). Formate is the only possible intermediate which would result in free energy gain, so thermodynamic models support formate as an electron shuttle (e-shuttle) of AOM (Sørensen et al., 2001). However, acetate was assumed to be the favorable e-shuttle in high $CH_4$ pressure environments (Valentine, 2002). Genomic studies suggested that the putative intermediates for AOM could be acetate, formate or hydrogen (Hallam et al., 2004; Meyerdierks et al., 2010). The formate dehydrogenase gene is highly expressed in the ANME-1 genome, thus formate can be formed by ANME-1 and function as intermediate (Meyerdierks et al., 2010). Likewise, the ADP-forming acetyl-CoA synthetase which converts acetyl-CoA to acetate was retrieved in the ANME-2a genome (Wang et al., 2014). Therefore, acetate could be formed by ANME-2a and be a possible intermediate. Considering AOM as a reversed methanogenesis, the first step is the conversion of $CH_4$ to methyl-CoM and the pathway involves the production of either acetate or hydrogen as an intermediate (Hallam et al., 2004; Wang et al., 2014). Nevertheless, the addition of hydrogen in an AOM experiment does not illustrate any change in AOM rate, in contrast to the typical methanogenesis process (Moran et al., 2008). Similarly, $CH_4$ based SR rates were the same even if these potential intermediates (acetate/formate) were supplied, whereas the reaction should be shifted to lower AOM rates upon the addition of intermediates (Meulepas et al., 2010a; Moran et al., 2008). Moreover, the addition of these potential intermediates induces the growth of different SRB than the *DSS* and *DBB* groups, which are the assumed syntrophic partner of ANME (Nauhaus et al., 2005). Therefore, the hypothesis of these compounds being possible AOM e-shuttles is unconfirmed. Instead, methyl sulfide was proposed to be an intermediate for both methanogenesis and methanotrophy (Moran et al., 2008). Methyl sulfide is then assumed to be produced by the ANME and can be utilized by the SRB partner (Moran et al., 2008).

However, few species are known to cooperate by direct electron transfer through conductive structures on the cell surfaces (Rotaru et al., 2014; Summers et al., 2010) Several mechanisms have been proposed for electron transfer: via microbial nanowires (Reguera et al., 2005), direct electron transfer via c-type cytochromes on the cell surfaces (Summers et al., 2010) or via conductive minerals (Kato et al., 2012) (Figure 5.2A). Multiheme *c*-type cytochromes were identified in the ANME-1 archaea genome (Meyerdierks et al., 2010) and the *c*-type cytochrome specific gene was also well expressed in the ANME-2a according to a metatranscriptome study (Wang et al., 2014). The importance of multiheme *c*-type cytochromes has been extensively discussed in *Geobacter* species, where the cytochrome can act as an electron storage in the cell membrane and subsequent extracellular e-transfer occurs (Lovley, 2008). These organisms use cell membrane cytochromes and pili as biological nanowires to connect between cell and mineral (Reguera et al., 2005). Recent studies gave some other evidence of the direct interspecies electron transfer between ANME and SRB showing a similar mechanism as for *Geobacter* (McGlynn et al., 2015; Wegener et al., 2015). Thermophilic ANME-1 and bacterial partners showed pili-like structures and they highly express genes for outer membrane c-type cytochromes (Wegener et al., 2015) and ANME-2 genome encodes large c-type cytochrome proteins (McGlynn et al., 2015).

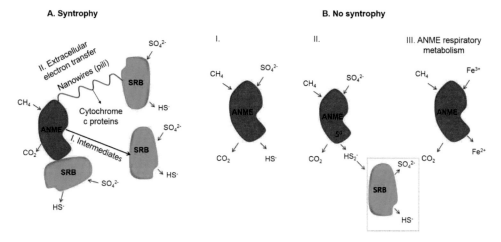

**Figure 2.5** Syntrophic and non-syntrophic ANME using sulfate as electron acceptor. A. In a syntrophic association ANME can transfer electron to SRB via different mechanisms: I) electron transfer via possible intermediate compounds such as, formate, acetate or hydrogen; II) electron transfer through cytochromes either via cell to cell contact between ANME and SRB or through biological nanowires such as pili (Scheller et al., 2016; Wegener et al., 2015). B. In a non-syntrophic association: I) ANME can possibly perform the complete AOM process alone without SRB or II) ANME can perform AOM by producing $CO_2$ and disulfide ($HS_2^-$) with $S^0$ as intermediate (Milucka et al., 2012); or III) ANME can be decoupled by SRB using an external electron acceptor (Scheller et al., 2016).

### 2.3.3 Non-syntrophic growth of ANME

Despite the recent discoveries about the cooperation between ANME and SRB the topic is still under debate. Visualization of ANME and its bacterial partners by FISH showed that for all three clades of ANME, the association with SRB is not obligatory. In some cases, the AOM process could occur by only the ANME without any $SO_4^{2-}$ reducing partner, especially for ANME-1 (Wankel et al., 2012a) and ANME-2 (Milucka et al., 2012) (Figure 2.5B). The possibility of non-syntrophic growth of ANME is further supported by the presence of nickel containing methyl-coenzyme M reductase (MCR) in ANME-1 and ANME-2, like other methanogens (Hallam et al., 2004; Wang et al., 2014). Scheller et al. (2010) discussed the MCR is able to break the stable C-H bond of $CH_4$ without any involvement of highly reactive oxidative intermediates.

Milucka et al. (2012) proposed a new AOM mechanism, in which ANME were responsible for both $CH_4$ oxidation and SR (Figure 2.5B). $CH_4$ was oxidized to bicarbonate and then the $SO_4^{2-}$ was reduced to zero-valent sulfur, as an intracellular intermediate in ANME-2 cells. The resulting sulfur was then released outside the cell as disulfide, which is converted to sulfide by the SRB. Figure 2.5B shows some ANME can sustain the overall AOM reaction without bacterial partner, even though the *DSS* type Deltaproteobacteria render the AOM-SR more thermodynamically favorable by scavenging the disulfide by disproportionation or dissimilatory reduction. The disulfide produced by ANME is disproportionated into $SO_4^{2-}$ and sulfide. The thus produced $SO_4^{2-}$ can be used again by the ANME, while sulfide can undergo several conversions, for instance precipitate as $FeS_2$ or partially oxidize (to $S^{o}$ ) or completely oxidize (to $SO_4^{2-}$) aerobically or anaerobically (in the presence of light by e.g. purple sulfur bacteria) (Dahl & Prange, 2006). As described earlier, in the presence of iron oxides, sulfide can react abiotically forming more substrates (disulfide and elemental sulfur) for the Deltaproteobacteria. The reaction of sulfide with iron oxides can thus strongly enhance the sulfur cycle, similarly to the study conducted with *Sulfurospirillum deleyianum* (Straub & Schink, 2004).

The cooperative/synergistic interaction between ANME and SRB is still unclear, as Milucka et al. (2012) stated that a syntrophic partner might not be needed for ANME-2, while recent studies have showed the interactions between the two partners by direct electron transfer (McGlynn et al., 2015; Wegener et al., 2016). However, Scheller et al. (2016) showed that ANME and SRB can be decoupled by using insoluble iron oxides as external electron acceptors and ANME is capable of respiratory metabolism (Figure 2.5B). Scheller et al. (2016) showed that ANME can live without the bacterial partner and thus with the possibility of growing ANME separately and fully understand the AOM mechanism may be possible in the future.

## 2.4    Drivers for the distribution of ANME in natural habitats

### 2.4.1    Major habitats of ANME

ANME are widely distributed in marine habitats including cold seep systems (gas leakage from $CH_4$ hydrates), hydrothermal vents (fissures releasing hot liquid and gas in the seafloor) and organic rich sediments with diffusive $CH_4$ formed by methanogenesis (Figures 2.6 and 2.7). The cold seep systems include mud volcanoes, hydrate mounds, carbonate deposits and gaseous carbonate chimneys (Boetius & Wenzhöfer, 2013), which are all frequently studied ANME habitats. The major controlling factors for the ANME distribution are the availability of $CH_4$ and $SO_4^{2-}$ or other terminal electron acceptors which can possibly support the anaerobic oxidation of $CH_4$, whilst other environmental parameters such as temperature, salinity, and alkalinity also play a decisive role in ANME occurrence. Among the three clades, ANME-2 and ANME-3 apparently inhabit cold seeps, whereas ANME-1 is cosmopolitan, residing in a wide temperature and salinity range (Table 2.1 and Figure 2.7). Recently, AOM has been reported in non-saline and terrestrial environments as well, for instance in the Apennine terrestrial mud volcanoes (Wrede et al., 2012) and in the Boreal peat soils of Alaska (Blazewicz et al., 2012) (Table 2.2).

The Black Sea, a distinct ANME habitat, consists of thick microbial mats of ANME-1 and ANME-2 (2-10 cm thick) adhered with carbonate deposits (chimney-like structure) in various water depths of 35-2000 m (Blumenberg et al., 2004; Michaelis et al., 2002; Novikova et al., 2015; Reitner et al., 2005; Thiel et al., 2001; Treude et al., 2005a). $CH_4$ is distributed by vein like capillaries throughout these carbonate chimneys and finally emanated to the water column (Krüger et al., 2008; Michaelis et al., 2002; Treude et al., 2005a). These microbial habitats are of different size and nature, such as small preliminary microbial nodules (Treude et al., 2005a), floating microbial mats (Krüger et al., 2008) and large chimneys (Michaelis et al., 2002). The immense carbonate chimney from the Black Sea (Figure 2.6A), with up to 4 m height and 1 m width, was found to harbor an ANME-1 dominant pink-colored microbial mat with the highest known AOM rates in natural systems (Blumenberg et al., 2004; Michaelis et al., 2002). Deep-sea carbonate deposits from cold seeps and hydrates are active and massive sites for AOM and ANME habitats (Marlow et al., 2014b). Likewise, $CH_4$ based authigenic carbonate nodules and $CH_4$ hydrates which host ANME-1 and ANME-2 (Marlow et al., 2014b; Mason et al., 2015; Orphan et al., 2001a; Orphan et al., 2001b; Orphan et al., 2002) prevail in the Eel River Basin (off shore California), a cold seep with an average temperature of 6°C and known for its gas hydrates (Brooks et al., 1991; Hinrichs et al., 1999). Both ANME-1 and ANME-2 are commonly associated with *DSS* in the sediments of Eel river, however ANME-1 appeared to exist as single filaments or monospecific aggregates in some sites as well (Hinrichs et al., 1999; Orphan et al., 2001b; Orphan et al., 2002).

Other cold seep sediments were also extensively studied as ANME habitats. The Gulf of Mexico, a cold seep with bottom water temperature of 6°C to 8°C, is known for its gas seepage and associated hydrates. These $CH_4$ hydrates located at around 500 m seawater depth in the Gulf of Mexico are inhabited by diverse microbial communities: *Beggiatoa* mats with active AOM are common bottom microbial biota in the sulfidic sediments (Joye et al., 2004; Lloyd et al., 2006; Orcutt et al., 2005; Orcutt et al., 2008). ANME-1 dominates the sediment of the Gulf of Mexico, particularly in the hypersaline part as a monospecific clade, whereas ANME-2 (a and b) are present together with *DSS* groups in the less saline hydrates (Lloyd et al., 2006; Orcutt et al., 2005). Similarly, different mud volcanoes of the Gulf of Cadiz cold seep harbor ANME-2 with the majority being ANME-2a (Niemann et al., 2006a), whereas the hypersaline Mercator Mud Volcano of the Gulf of Cadiz hosts ANME-1 (Maignien et al., 2013). Retrieval of ANME-1 in the hypersaline environment suggests the ANME-1 adaptability to wider salinity ranges compared to other ANME phylotypes. Mud volcanoes from the Eastern Mediterranean (Kazan and Anaximander mountains) are inhabited by all three ANME phylotypes, whereas Kazan Mud Volcano hosts the distinct ANME-2c clade (Heijs et al., 2007; Kormas et al., 2008; Pachiadaki et al., 2010; Pachiadaki et al., 2011). Likewise, Haakon Mosby Mud Volcano (HMMV) in the Barents Sea is the firstly described habitat for ANME-3 with almost 80 % of the microbial cells being ANME-3 and *DBB* (Figures 2.6B and 2.2D) (Losekann et al., 2007; Niemann et al., 2006b).

Some of the hydrothermal vents are well studied ANME habitats for distinct ANME clades and thermophilic AOM. The Guaymas Basin in the California Bay, an active hydrothermal vent with a wide temperature range, is known for the occurrence of different ANME-1 phylotypes, along with unique thermophilic ANME-1 (Biddle et al., 2012; Larowe et al., 2008; Vigneron et al., 2013). ANME-1 is predominant throughout the Guaymas Basin, yet the colder $CH_4$ seeps of the Sonara Margin host all three ANME phylotypes (ANME-1, ANME-2 and ANME-3) with peculiar ANME-2 (ANME-2c Sonara) (Vigneron et al., 2013). Likewise, mesophilic to thermophilic AOM carried out by the ANME-1 clade was detected in the Middle Valley vent field on the Juan de Fuca Ridge (Lever et al., 2013; Wankel et al., 2012a). Another vent site, the Lost City hydrothermal vent with massive fluid circulation and ejecting hydrothermal fluid of >80°C predominantly hosts ANME-1 within the calcium carbonate chimneys (Figure 2.6C), which are very likely deposited due to bicarbonate formation from AOM (Bradley et al., 2009; Brazelton et al., 2006).

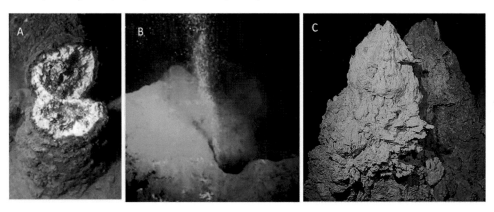

**Figure 2.6** *In situ* pictures of some of the well studied ANME habitas. A) Giant microbial mat in carbonate chimney in the Blak Sea (Blumenberg et al., 2004), B) $CH_4$ bubble seeping from Haakon Mosby mud volcano and C) carbonate chimney from the Lost City hydrothermal vent (Brazelton et al., 2006).

### 2.4.2 ANME types distribution by temperature

The ANME clades exhibit a distinct pattern of distribution according to the temperature. ANME-2 and ANME-3 seem more abundant in cold seep environments, including hydrates and mud volcanoes, with an average temperature of 2 to 15°C. In contrast, ANME- 1 is more adapted to a wide temperature range from thermophilic conditions (50-70°C) to cold seep microbial mats and sediments (4-10°C) (Holler et al., 2011b; Orphan et al., 2004). Temperature appears to control the abundance of the ANME clades. However, some of the ANME types (ANME-1ab) exhibit adaptability to a wide range of temperatures. Other geochemical parameters as salinity, $CH_4$ concentration and pressure can act together with temperature as selection parameters for the distribution of ANME in natural environments.

ANME-1 was extensively retrieved across the temperature gradient between 2°C to 100°C in the Guaymas Basin from the surface to deep sediments (Teske et al., 2002). A phylogenetically distinct and deeply branched group of the ANME-1 (ANME-1GBa) was found in the high temperature Guaymas Basin hydrothermal vent (Biddle et al., 2012) and other geologically diverse marine hydrothermal vents such as the diffuse hydrothermal vents in Juan de Fuca Ridge in the Pacific Ocean (10-25 °C) (Merkel et al., 2013). The thermophilic trait of ANME-1GBa is supported by its GC (guanine and cytosine) content in its 16S rRNA genes, as it holds a higher GC percentage (>60 mol %) compared to other ANME types. The GC content is positively correlated with the optimum temperature of microbial growth, the elevated GC content of ANME-1GBa suggests ANME-1 GBa being a thermophilic microbial cluster, with on optimum growth temperature of 70°C or above (Merkel et al., 2013). Moreover, when the Guaymas ANME community was enriched *in vitro*, the highest AOM rate was obtained in the range of 45-60°C, indicating that the major community consists of thermophilic ANME-1 (Holler et al., 2011b).

Other ANME-1 phylotypes (ANME-1a and ANME-1b) were observed in wide temperature ranges (3°C to > 60°C) (Biddle et al., 2012). ANME-1a and ANME-1b were retrieved from different hydrothermal vent areas and cold seeps, for example the Guaymas Basin hydrothermal vent at >60°C (Biddle et al., 2012), Lost City hydrothermal vent (Brazelton et al., 2006), the Sonora Margin cold seep of the Guaymas Basin (3°C) (Vigneron et al., 2013), mud volcanoes in the Eastern Mediterranean cold seep (14-20°C) (Lazar et al., 2012), the Gulf of Mexico (6°C) (Lanoil et al., 2001; Lloyd et al., 2006), Black Sea microbial mat and water column (8°C) (Knittel et al., 2005; Schubert et al., 2006) and Eel River Basin (6°C) (Hinrichs et al., 1999; Orphan et al., 2001b). The occurrence of ANME-1a and ANME-1b in cold seep environments suggests ANME-1a and ANME-1b to be putative mesophiles to psychrophiles. The GC percentage of 16S rRNA genes of ANME-1a and ANME-1b is around 55 mol %, which is common for mesophiles (Merkel et al., 2013).

In contrast, ANME-2 and ANME-3 have a narrow temperature range. ANME-2 clades (2a, 2b and 2c) appear predominant in marine cold seeps and in some SMTZs where the temperature is about 4-20°C. The major cold seep environments inhabited by ANME-2 are described in the previous section (section 2.4.1). The adaptability of ANME-2 in the cold temperature range is also substantiated by bioreactor enrichments with Eckernförde Bay sediment, where the maximum AOM rate was obtained when the bioreactor was operated at 15°C rather than at 30°C, for ANME-2a (Meulepas et al., 2009a). Similarly, Eckernförde Bay *in vitro* AOM rate measurements showed a steady increment in AOM rates from 4°C to 20°C and subsequently decreased afterwards (Treude et al., 2005b). Conversely, the recently described clade ANME-2d affiliated "*Candidatus Methanoperedens nitroreducens*" (Figure 2.3C), which was enriched from a mixture of freshwater sediment and wastewater sludge (Haroon et al., 2013), grows optimally at mesophilic temperatures (22-35°C) (Hu et al., 2009).

ANME-3 is also known to be thriving in cold temperature environments including cold seeps and mud volcanoes. The ANME-3 clade was firstly retrieved from the Haakon Mosby Mud Volcano with a temperature of about -1.5 °C (Niemann et al., 2006b). Later, ANME-3 was found in other cold seep areas as well, such as the Eastern Mediterranean seepages at about 14°C (Heijs et al., 2007; Pachiadaki et al., 2010) and the Skagerrak seep (Denmark, North Sea) at around 6-10°C (Parkes et al., 2007).

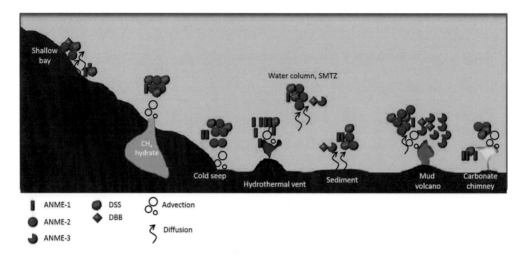

**Figure 2.7** Major habitats of ANME in marine environments and ANME distribution along the different major habitats. ANME-1 is mainly inhabited in diverse environments including hydrothermal vents, cold seeps and carbonates chimneys, whereas ANME-2 was retrieved from diverse cold seeps, $CH_4$ hydrates and mud volcanoes and ANME-3 was mainly retrieved from a specific mud volcano. ANME types: (❚) ANME-1, (●) ANME-2 and (◗) ANME-3. SRB types: *DSS* ( ● ) and *DBB* (◆). $CH_4$ transport regime: advection ( O ) and diffusion ( ⌇ ).

### 2.4.3  CH₄ supply mode as driver for distribution of ANME

In some seafloor ecosystems, $CH_4$ is transported by diffusion due to concentration gradients. Diffusion dominated ecosystems are typically quiescent sediments. In contrast, in seafloor ecosystems with $CH_4$ seeps, $CH_4$ is transported by advection of $CH_4$-rich fluids. Due to the complex dynamics of $CH_4$ transport in advection dominated environments, estimations of *in situ* $CH_4$ oxidation rates by geochemical mass balances is rather difficult (Alperin & Hoehler, 2010). Based on *ex situ* tests, the AOM rates are higher in ecosystems where high CH4 fluxes are sustained by advective transport than in diffusion dominated ecosystems (Boetius & Wenzhöfer, 2013). The velocity of the $CH_4$-rich fluid may result in an order of magnitude difference in AOM rates. Higher AOM rates were observed at sites with higher flow velocity (Krause et al., 2014), probably high flows of $CH_4$-rich fluid support dense ANME populations.

The extent of $CH_4$ flux and the mode of $CH_4$ transport (advection vs diffusion) are certainly important drivers for ANME population dynamics. Mathematical simulations illustrate that the transport regime can control the activity and abundance of AOM communities (Dale et al., 2008). We performed multivariate and cluster analysis with data from the literature showing the mode of $CH_4$ transport can possibly control AOM communities (Figure 2.8).

**Table 2.1** Rates of AOM and SR in different natural marine habitats along with dominant ANME types. Here the depth refers to water depth. The different methods of AOM and SR measurements are indicated by the superscript letters next to the references as follows: [a]= *in vitro* measurement, [b]= *ex situ* radiotracer measurement, [c]= model calculation, [d]= pore water chemistry measurement

| Location | Depth (m) | CH₄ (mM) | SO₄²⁻ (mM) | ANME types | AOM rates | SR rates | References |
|---|---|---|---|---|---|---|---|
| **Cold seeps (temperature ranging from 1.5 to 20°C)** | | | | | | | |
| Black Sea (giant carbonate chimney) | 230 | 2.8 | 17 | ANME-1 | 7800 to 21000 nmol l⁻¹ g$_{dw}$⁻¹ d⁻¹ | 4300 to 19000 µmol g$_{dw}$ ⁻¹ d⁻¹ | (Michaelis et al., 2002; Treude et al., 2007)[a] |
| Black Sea (other microbial mats) | 180 | 3.7 | 9 to 15 | ANME-1, ANME-2 | 2000 to 15000 nmol g$_{dw}$ ⁻¹ d⁻¹ | 4000 to 20000 nmol g$_{dw}$ ⁻¹ d⁻¹ | (Krüger et al., 2008)[a] |
| Haakon Mosby mud volcano, Barents Sea | 1250 | 0.0003 to 0.0057 | - | ANME-2, ANME-3 | 1233 to 2000 nmol cm⁻² d⁻¹ | 2250 nmol cm⁻² d⁻¹ | (Niemann et al., 2006b) [b] |
| Gulf of Mexico, hydrate | 550 to 650 | 2 to 6 | 20 | ANME-1, ANME-2 | 280 ± 460 nmol cm⁻² d⁻¹ | 5400 ± 9400 nmol cm⁻² d⁻¹ | (Joye et al., 2004; Orcutt et al., 2005) |
| Eel River Basin carbonate mounds and hydrates | 500 to 850 | 3 | 20 | ANME-1, ANME-2 | 200 nmol cm⁻³ d⁻¹ | - | (Marlow et al., 2014b; Orphan et al., 2004)[b] |
| Gulf of Cadiz, mud volcanoes | 810 to 3090 | 0.001to 1.3 | 10 to 40 | ANME-2, ANME-1 | 10 to 104 nmol cm⁻² d⁻¹ | 158 to 189 nmol cm⁻² d⁻¹ | (Niemann et al., 2006a)[b] |
| Black Sea water | 100 to 1500 | 0.011 | - | ANME-, ANME-2 | 0.03 to 3.1 nmol d⁻¹ | - | (Durisch-Kaiser et al., 2005; Schubert et al., 2006)[b] |
| Tommeliten seepage area, North Sea sediment | 75 | 1.4 to 2.5 | 30 to 20 | ANME-1, ANME-2 | 1.4 to 3 nmol cm⁻³ d⁻¹ | 3 to 4.6 nmol cm⁻³ d⁻¹ | (Niemann et al., 2005)[b] |

**CH₄ rich sediments (temperature from 4 to 20°C)**

| | | | | | | | |
|---|---|---|---|---|---|---|---|
| Bothnian Sea sediment | 200 | 2 | 5.5 | - | 40 -90 nmol cm$^{-2}$ d$^{-1}$ | - | (Slomp et al., 2013)[c] |
| Baltic Sea/ Eckernförde Bay sediment | 25 | 0.001 to 0.8 | 16 to 21 | ANME-2 | 1 to 14 Nmol cm$^{-3}$ d$^{-1}$ | 20 to 465 nmol cm$^{-3}$ d$^{-1}$ | (Treude et al., 2005b)[b] |
| Skagerrak sediment | 308 | 1.3 | 25 | ANME-2 and ANME-3 | 3 nmol cm$^{-3}$ d$^{-1}$ | - | (Parkes et al., 2007)[d] |
| West African margin sediment | 400 to 2200 | 1 to 19 | 26 | - | 0.0027 nmol cm$^{-3}$ d$^{-1}$ | - | (Sivan et al., 2007)[c] |

**Hydrothermal vents (temperature from 10 to 100°C)**

| | | | | | | | |
|---|---|---|---|---|---|---|---|
| Guaymas Basin hydrothermal vent | - | - | - | ANME-1 | 1200 nmol g$_{dw}$$^{-1}$ d$^{-1}$ | 250 nmol g$_{dw}$$^{-1}$ d$^{-1}$ | (Holler et al., 2011b)[a] |
| Juan de Fuca Ridge hydrothermal vent | 2400 | 3 | - | ANME-1 | 11.1 to 51.2 nmol cm$^{-3}$ d$^{-1}$ | - | (Wankel et al., 2012a)[c] |

CH$_4$-rich upward fluid flow at active seep systems restricts AOM to a narrow subsurface reaction zone and sustains high CH$_4$ oxidation rates. In contrast, pore-water CH$_4$ transport dominated by molecular diffusion leads to deeper and broader AOM zones, which are characterized by much lower rates and biomass concentrations (Dale et al., 2008). In this context, Roalkvam et al. (2012) found that the CH$_4$ flux largely influenced the specific density of ANME populations. However, whether distinct ANME types preferentially inhabit environments dominated by advective or diffusive CH$_4$ transport is not yet clear. At sites with high seepage activity like the Hydrate Ridge in Oregon, ANME-2 was dominant, whereas ANME-1 apparently was more abundant in the low seepage locations (Marlow et al., 2014a). A rough estimate of the abundance of the ANME type populations, reported in various marine environments, shows that ANME-2 dominate sites where CH$_4$ is transported by advection, while ANME-1 may dominate sites where CH$_4$ is transported by diffusion or advection (Figure 2.8). It is advisable that future studies regarding ANME type's distribution explicitly indicates the dominant mode of *in situ* CH$_4$ transport.

**Table 2.2** Rates of AOM and SR in different natural terrestrial habitats. The different methods of AOM and SR measurements are indicated by the superscript letter next to the reference as follows: [a]= *in vitro* measurement, [b]= *ex situ* radiotracer measurement

| Location | Soil depth (cm) | CH$_4$ (mM) | SO$_4^{2-}$ (mM) | AOM rates | SR rates | References |
|---|---|---|---|---|---|---|
| Wetland and peat soil | 0-40 | 0.5-1 | 0.1-1 | 265 ± 9 nmol cm$^{-3}$ d$^{-1}$ | 300 nmol cm$^3$d$^{-1}$ | (Segarra et al., 2015)[b] |
| Tropical forest soil of Alaska | 10-15 | - | <1 | 3-21 nmol g$_{dw}^{-1}$ d$^{-1}$ | - | (Blazewicz et al., 2012)[a] |
| Paclele Mici Mud Volcano in Carpathian mountains | - | - | 1.5-2 | 2-4 nmol g$_{dw}^{-1}$d$^{-1}$ | ANME-2 | (Alain et al., 2006)[a] |
| Peat land soil from diverse places | 30-50 | | - | 0.03-3 nmol g$_{dw}^{-1}$ d$^{-1}$ | - | (Gauthier et al., 2015)[a] |

## 2.5   *Ex situ* enrichment of ANME

### 2.5.1   Need for enrichment of ANME
Molecular based methods allow the recognition of the phylogenetic diversity of ANME microorganisms in a wide range of marine sediments and natural environments. Determination of their detailed physiological and kinetic capabilities requires, until now, the cultivation and isolation of the microorganisms. The culturability of microorganisms inhabiting seawater (0.001-0.1%), seafloor (0.00001-0.6%) and deep subsea (0.1%) sediments is among the lowest compared to other ecosystems (Amann et al., 1995; D'Hondt et al., 2004; Parkes et al., 2000).

This also holds for ANME from all thus far known environments, which so far have not yet been cultivated in pure culture for various reasons, not all known.

Specifically for the enrichment of ANME, the following aspects limit their cultivation: (i) from all known microbial processes, the AOM reaction with $SO_4^{2-}$ is among those which yield the lowest energy, (ii) the growth rate of these microorganisms is thus very low with a yield of 0.6 $g_{celldw}$ per mol of $CH_4$ oxidized (Nauhaus et al., 2007), (iii) the dissolved concentrations of their substrate $CH_4$ (1.4 mM) at atmospheric pressures is limited to values far much lower than the estimated apparent half affinity constant for $CH_4$ (37 mM) during the AOM process and (iv) sulfide, which is a product of the reaction, can be inhibitory. All these aspects set a great challenge for the cultivation and isolation of ANME.

It is recognized that culturability can be enhanced when the conditions used for cultivation mimic well those of the natural environment. Cultivation efforts have been focused mainly on increasing dissolved $CH_4$ concentrations. To enrich AOM *ex situ*, batch and continuous reactors operated at moderate and high pressures have been tested. To avoid potential sulfide toxicity, attention has been paid to exchange the medium so that the sulfide concentrations do not exceed 10 to 14 mM (Nauhaus et al., 2007; Nauhaus et al., 2002).

### 2.5.2   Conventional *in vitro* ANME enrichment techniques

The conventional *in vitro* incubation in gas tight serum bottles provides an opportunity to test the microbial activities, kinetics of the metabolic reactions and the enrichment of the microbes, more specifically for the large number of uncultured anaerobes like ANME. Conventional serum-bottles are widely used when the incubation pressures do not exceed 0.25 MPa (Beal et al., 2009; Blumenberg et al., 2005; Holler et al., 2011a; Meulepas et al., 2009b).

A batch bottle experiment provides the flexibility to operate many different experiments in parallel (large numbers of experimental bottles can be handled at the same time) by controlling different environmental conditions such as temperature, salinity or alkalinity. The batch incubation based experiments are relatively easy to control and manipulate, especially with very slow growing microbes like ANME, which require strictly anaerobic conditions. AOM activity is negligible in the presence of oxygen (Treude et al., 2005b). The commonly used batch serum bottles or culture tubes with thick butyl rubber septa facilitate the sampling while maintaining the redox inside, although there are several other factors which can be key for ANME enrichment, such as the low solubility of $CH_4$ and the possible accumulation of sulfide toxicity in the stationary batches.

As shown in Table 2.1 and Table 2.3, several studies estimated the AOM rate by *in vitro* batch incubations (Holler et al., 2011b; Kruger et al., 2008; Wegener et al., 2008). Kruger *et al.* (Krüger et al., 2008) determined AOM rates from 4000 to 20000 nmol $g_{dw}^{-1}$ day$^{-1}$ by incubating microbial mats from the Black Sea. Holler et al. (2011a) estimated AOM at a rate of 250 nmol $g_{dw}^{-1}$ day$^{-1}$ by ANME-1 from the Black Sea (Table 2.1). ANME-2 dominated communities from the Hydrate Ridge of northeast Pacific exhibit 20 times higher specific AOM rates (20 mmol day$^{-1}$ $g_{dw}^{-1}$) compared to ANME-1 from the Black Sea pink microbial mat (Nauhaus et al., 2005). During the *in vitro* incubations with different environmental conditions, unlike the

$SO_4^{2-}$ concentration, pH and salinity variations, temperature was found to be a major influential parameter for AOM rates in ANME-1 and ANME-2 communities (Nauhaus et al., 2005). Both ANME communities showed the increment in AOM rate with elevated $CH_4$ partial pressure. However, when the microbial mat from the Black Sea with both ANME-1 and ANME-2 was incubated in batch at low $CH_4$ concentrations, ANME-1 growth was favored over the growth of ANME-2 (Blumenberg et al., 2005).

Optimum pH, temperature, salinity and sulfide toxicity were determined as 7.5, 20°C, 30 ‰ and 2.5 mM, respectively, for the ANME-2 enrichment from Eckernförde Bay when incubated in 35 ml serum bottles (Meulepas et al., 2009b). The highest *in vitro* AOM activity was obtained at 15°C compared to other temperature incubations (Treude et al., 2005b) and sulfide toxicity was reported beyond 2.5 mM for Eckernförde Bay sediments (Meulepas et al., 2009b). Likewise, possible electron donors and acceptors involved in the AOM process were studied in the batch incubations. The sediment from Eckernförde Bay was incubated with different methanogenic substrates for the study of possible intermediates between the ANME and SRB (Meulepas et al., 2010a). The AOM activity with other electron acceptors than $SO_4^{2-}$, i.e. Fe (III) and Mn (IV), by Eel river sediment was estimated by batch incubations for the detection of iron/manganese dependent AOM (Beal et al., 2009). Moreover, thermophilic AOM was studied in batch assays within different temperature ranges (up to 100°C) with Guaymas Basin hydrothermal vent sediment, AOM was observed up to 75°C with the highest AOM rate at 50°C (Holler et al., 2011b).

### 2.5.3 Modified *in vitro* ANME enrichment approaches

The growth of ANME-2 was documented (Nauhaus et al., 2007) in batch incubations using a glass tube connected via a needle to a syringe and placed inside a pressure-proof steel cylinder (Nauhaus et al., 2002). The syringe, which is filled with medium, transmits the pressure of the cylinder to the medium inside the tube. Using this design, $CH_4$ hydrate sediment was incubated at 1.4 MPa for 2 years with intermittent replenishment of the supernatant by fresh medium and $CH_4$ (21 mM at 12°C). During the incubation period, the volume of the ANME-2 and SRB consortia, which was tracked using FISH, increased exponentially (Nauhaus et al., 2007).

A batch incubation with intermittent replacement of supernatant by fresh medium (i.e., fed batch system) once a month was used to successfully enrich ANME-2d at abundances of about 78% (Haroon et al., 2013). The inoculum was a mixture of sediment from a local freshwater lake, anaerobic digester sludge and activated sludge from a wastewater treatment plant in Brisbane, Australia (Table 2.3) (Hu et al., 2009).

The retention of biomass in the fed-batch system was achieved via a 20 min settling period prior to the replacement of the supernatant by fresh medium. The cultivation of this freshwater ANME-2d can have the advantage of higher solubility of $CH_4$ in freshwater than in seawater (Yamamoto et al., 1976), however, this microorganism was enriched at 35 °C and $CH_4$ solubility decreases at increased temperatures (Hu et al., 2009). As previously specified (section 2.2.5), this ANME-2d, named *"Candidatus Methanoperedens nitroreducens"*, utilizes nitrate instead of $SO_4^{2-}$ as electron acceptor for AOM. This physiological trait likely contributed to the successful enrichment of this novel ANME clade at high abundance in a relatively short time period (about 2 years), because AOM coupled to nitrate yields about 45-fold more energy than its $SO_4^{2-}$ dependent counterpart (Figure 2.4).

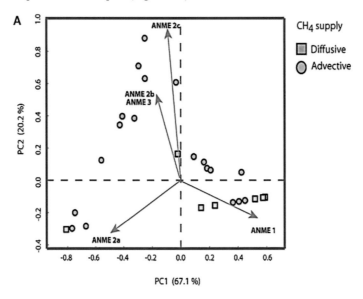

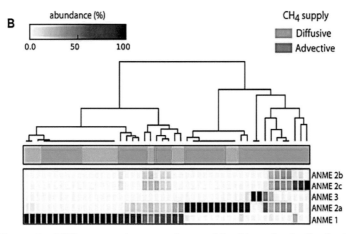

**Figure 2.8** The mode of $CH_4$ transport is apparently one of the drivers for the distribution of ANME types in the environment. A) Multivariate and B) cluster analyses show that ANME-2 is dominant mostly in $CH_4$-advective sites.

### 2.5.4   Continuous bioreactor based ANME enrichment

The design rationale of continuous flow incubation columns is to provide nutrients and to remove end products at environmentally relevant rates (Table 2.3) (Girguis et al., 2003). In such systems, 0.2 μm filtered seawater, reduced with hydrogen sulfide (510 μM) and saturated with $CH_4$ (1.5 mM) in a conditioning column (4 h at 0.5 MPa), was used to feed cold seep and non-seep sediment cores maintained in PVC tubes at 0.2 MPa and 5°C (Girguis et al., 2003). The $CH_4$ oxidation rates before and after incubations of the seep sediments were the same, probably because the incubation time was only 2 weeks. However, increase in AOM rate and ANME-2c population size were detected in the non-seep sediment incubations. In a second experimental run, the same continuous flow reactor was used, but the incubations were conducted at 1 MPa. The incubation time was 7.5 months (30 weeks) and a preferential proliferation of ANME-1 against ANME-2 was observed in the non-seep sediments at the highest pore water velocity tested (90 m year$^{-1}$) (Girguis et al., 2005). In addition, an increase in the AOM activity was reported as measured using batch incubations in serum bottles inoculated by the sediment (seep and non seep sediments used in the continuous enrichment experiment) without headspace, using 0.2 μm filter-sterilized anoxic seawater containing 2.0 mM $CH_4$ and 1 mM hydrogen sulfide (Girguis et al., 2005).

In efforts to attain $CH_4$ concentrations close to *in situ* values, continuous reactors that can handle hydrostatic pressures up to 44.5 MPa with $CH_4$ enriched medium and without free gas in the incubation chamber have been used (Deusner et al., 2009). This reactor configuration is flexible to operate in batch, fed-batch or continuous mode. Incubation of sediments from the Black Sea showed a six-fold increase in the volumetric AOM rate when the $CH_4$ partial pressure increased from 0.2 to 6 MPa. In all operation modes, AOM rates were estimated based on sulfide production. However, when in otherwise similar operation conditions $CH_4$-saturated medium was replaced by $CH_4$-free medium, sulfide levels decreased rapidly and stabilized at input levels. This indicated that the sulfide production was indeed coupled to $CH_4$ oxidation. During continuous operation of such high pressure reactors, a $CH_4$ concentration of 60-65 mM can be readily attained. Noticeably, during continuous operation, the influent $SO_4^{2-}$ concentration used was 8 mM, which is lower than seawater concentrations (Deusner et al., 2009). The hydraulic retention time was set at 14 h which corresponded to a dilution rate of 1.7 day$^{-1}$. Assuming a completely mixed reactor, this means that microorganisms growing at rates < 1.7 day$^{-1}$ would be washed-out from the reactor, which is the case of ANME having much lower growth rates (0.006 to 0.03 day$^{-1}$ (Girguis et al., 2005; Meulepas et al., 2009a; Meulepas et al., 2009b). Additionally, these tests of continuous operation with $CH_4$ addition lasted only 16 days and whether and how biomass was retained in the system was not reported (Deusner et al., 2009).

Similar high pressure systems have been operated at up to 60 MPa hydrostatic pressure and 120°C (Sauer et al., 2012). The flexibility of this system allows the sub-sampling of medium without loss of pressure and it can be operated in batch or continuous mode (Sauer et al., 2012). The system was tested incubating sediments from the Isis Mud volcano from the Egyptian continental margin (~ 991 m below sea level) using artificial seawater pre-conditioned with 4 MPa of $CH_4$ resulting in dissolved concentrations of ~ 96 mM $CH_4$. Following $CH_4$ saturation, the hydrostatic pressure was increased to 10 MPa using artificial seawater and incubations were conducted for 9 days at 23°C. No measurements of biomass concentration and yield were conducted, but an increase in sulfide was detected upon addition of $CH_4$ to the reactor (Sauer et al., 2012).

A continuous high pressure reactor capable to withstand up to 8 MPa was used in fed-batch and continuous mode at pressures from 1 to 8 MPa and a hydraulic retention time of 100 h during a 286 days incubation of sediments from a mud volcano located in the Gulf of Cadiz (Zhang et al., 2010). Under such conditions, the ANME-2 biovolume (count of cells and aggregates) increased about 12-fold (Zhang et al., 2011). There was no indication about the biomass retention time and AOM rate in the system.

ANME can also be enriched at moderate pressures or even ambient pressure provided biomass retention is applied. The latter can be achieved by introducing a submerged membrane (pore size 0.2 μm and effective surface of 0.03 $m^2$) within the reactor (Meulepas et al., 2009a). $CH_4$ was sparged continuously at 190 mmol $l^{-1}$ $day^{-1}$, thus providing mixing, stripping-off of the sulfide and restricting fouling of the membrane. This bioreactor was operated at 15°C and at a slight over pressure (0.25 MPa) to avoid $O_2$ intrusion. The $SO_4^{2-}$ loading rate was 3 mmol $l^{-1}$ $day^{-1}$ and the hydraulic retention time 7 days. Sediment retrieved from the Eckernförde Bay in the Baltic Sea was used as inoculum and the reactor was operated for about 3 years. Growth of ANME was inferred by the increase in sulfide production in the membrane reactor, and the increase in AOM rates was monitored by carrying out batch experiment with reactor biomass amended with [13]C-labelled $CH_4$ at regular time intervals (Meulepas et al., 2009a). The ANMEs in the reactor could be affiliated to ANME-2a and their doubling time was estimated at 3.8 month (i.e., growth rate 0.006 $day^{-1}$).

Although high pressure reactors operate at high dissolved $CH_4$ concentrations, their maintenance and operation is cumbersome and requires meeting various safety criteria for their implementation. When successful enrichment has been reported at moderate pressures in fed-batch reactors, a key feature was a good biomass retention via settling (ANME-2d) (Haroon et al., 2013) or membranes (ANME-2c) (Meulepas et al., 2009a).

## 2.5.5   Future development in *ex situ* enrichment approaches

Mimicking the natural conditions in bioreactors can be a fruitful strategy for enrichment of ANME. Reproducing *in situ* conditions in the laboratory is quite challenging, but artificial material and equipment can be used to mimic the natural environment (Figure 2.9). Mimicking natural conditions is possible by using suitable reactors capable of achieving extreme environmental conditions such as high pressure or temperature and with suitable or similar natural packing material. The carbonate-minerals, where ANME have been found to form microbial reefs, are very porous. This porous natural matrix can harbor aggregates of AOM performing consortia (Marlow et al., 2014b). Similarly, polyurethane sponges are a porous material and can be used as packing material in a packed bed bioreactor configuration to promote the adhesion, aggregation and retention of biomass. The collected marine sediment can be entrapped in the porous sponges so that $CH_4$ can effectively diffuse through them, while the medium containing necessary nutrients and electron acceptor flows through the material (Imachi et al., 2011). In a recent study, fresh bituminous coal and sandstone collected from a coal mine were used in a flow through type reactor system at high pressure to simulate and study geological $CO_2$ sequestration and transformation (Ohtomo et al., 2013). Similarly, the naturally occurring materials can be used as packing materials in bioreactors which assist in biomass retention in the ANME enrichment bioreactor.

Considering the importance of substrate availability, especially for ANME which are oxidizing a poorly soluble compound like $CH_4$, membrane reactors can be used to facilitate the contact between substrate and biomass. A hollow-fiber membrane reactor was successfully applied for $CH_4$-dependent denitrification (Shi et al., 2013). $CH_4$ passes internally the hollow-fiber membranes and diffuses to the outside layer where a biofilm of ANME can be retained and grown (Figure 2.9). A silicone membrane can also be used as a hollow-fiber membrane, which allows bubbleless addition of gas to the bioreactor compartment. These gas diffusive membranes are also applicable for AOM coupled to $SO_4^{2-}$ reduction, where the diffused $CH_4$ can be immediately taken up by ANME consortia which are suspended in the $SO_4^{2-}$ containing medium.

This mode of $CH_4$ supply produces minimum bubbles and the gas supply can be controlled by maintaining the gas pressure inside the membrane. As the microbial metabolism of AOM is slow, the slow diffusion of $CH_4$ can reduce the large amounts of unused $CH_4$ released from a bioreactor system, thus reducing the operational costs. Another benefit of the membrane is the biomass retention, as the biomass usually develops as biofilm or flocs (Jagersma et al., 2009). Moreover, the sulfide and pH can be continuously monitored by using pH and pS (sulfide sensor) electrodes and the sulfide can be removed before reaching the toxic threshold. A process control algorithm has been developed for the SR process, which is also applicable for AOM studies (Cassidy et al., 2015).

Several studies hypothesized an electron transfer between ANME and SRB (section 2.3). Based on this assumption, bio-electrochemical systems (BES) could also be used to study electron transfer mechanisms. The $CH_4$ oxidation process by the ANME takes place at the anode and SR takes place at the cathode (Figure 2.9). Using BES, compounds, which can act as e-shuttles (e.g. electron mediators or conductive nanominerals such as iron oxides) between the electrodes and ANME, can be added to facilitate the electron transfer from the ANME to the electrode and study the mechanism of electronic communication (Rabaey & Rozendal, 2010). The electron exchange between the electrodes and the ANME can be determined by applying different electrode potentials (Lovley, 2012). Another advantage of AOM studies using BES is to isolate or enrich the ANME. Assuming that the bacterial partner is required, a conductive membrane or electrode as electron sink can be used, which can act as bacterial partner and overcome its requirement. In addition, the electrode can be poised at a desired potential to serve as e-acceptor, then ANME growth can possibly be maximized by fine-tunig the electrode potential. Thus, the electrodes in BES facilitate experiments with electron transfer of $CH_4$ to the conducting surface and also serve as e-acceptor by which the ANME growth is possibly accelerated.

**Table 2.3** Enrichment condition and AOM rate for the *in vitro* studies of AOM. Here, incubation temperature, pressure, ANME growth rate and apparent affinity is represented by T, p, µ and Km, respectively. DHS refers to downflow hanging sponge bioreactor. Different mineral medium used for incubation are indicated by the superscript letters next to each reference. [a] = the incubation in artificial salt water mineral medium prepared according to Widdel and Bak (1992), [b] = the incubation with filter sterilized sea water and [c] = incubation in fresh water medium with nitrate and ammonium. SR represents for $SO_4^{2-}$ reduction.

| Enrichment mode | Inocula and incubation period (d) | T (°C) | p (MPa) | ANME types | AOM rate ($\mu mol\ g_{dw}^{-1}\ d^{-1}$) | ANME doubling time (months) | µ ($d^{-1}$) | Km (mM) | References |
|---|---|---|---|---|---|---|---|---|---|
| Fed-batch | AOM and Anamox, 230-290 d | 22-35 | 0.05 - 0.1 | ANME-2d | 1100 $\mu M\ d^{-1}$ | - | - | - | (Haroon et al., 2013)[c] |
| Membrane bioreactor, continuous well mixed | Baltic Sea/ Eckernförde Bay, 884 d | 15 | 0.1 | ANME-2a | 286 | 3.8 | 0.006 | <0.5 mM (for $SO_4^{2-}$) 0.075 MPa for $CH_4$ | (Meulepas et al., 2009a; Meulepas et al., 2009b)[a] |
| Fed-batch | Hydrate Ridge, North- east Pacific, 700 d | 15 | 1.4 | ANME-2 | 230 | 7 | 0.003 | >10mM ($CH_4$) | (Nauhaus et al., 2007)[a] |
| Batch | Gulf of Mexico, 150 d | 12 | 1.5 | ANME-1 | 13.5 | 2 | - | - | (Kruger et al., 2008)[a] |
| Fed-batch | Gulf of Cadiz, 286 d | 15 | 8.0 | ANME-2 | 9.22 (SR) | 2.5 | - | 37 mM ($CH_4$) | (Zhang et al., 2010; Zhang et al., 2011)[a] |
| Batch | Guaymas Basin sediment, 250 d | 42 - 65 | 0.25 | ANME-1 | 1.2 | 2.3 | - | - | (Holler et al., 2011b)[a] |
| DHS bioreactor | Nankai Trough, 2013 d | 10 | 0.1 | ANME-1,-2,-3 | 0.375 | - | - | - | (Aoki et al., 2014)[a] |
| Anaerobic $CH_4$ incubator system (continuous) | Monterey Bay, 400 d | 5 | 0.1 | ANME-1 ANME-2 | $9\times10^{-3}$ (ANME-1), 0.138 (ANME-2) | 1.1 (ANME-2) 1.4 (ANME-1) | 0.03 (ANME-1) 0.024 (ANME-2) | - | (Girguis et al., 2005)[b] |

## 2.6 Approaches for AOM and ANME studies

### 2.6.1 Measurement of AOM rates in activity tests

Various geochemical and microbial analyses are carried out for ANME and AOM studies. The common approach used to identify the occurrence of AOM is by direct $CH_4$, $CO_2$, $SO_4^{2-}$ and $S^{2-}$ profile measurements in marine environments and batch incubations (Reeburgh, 2007) *and reference therein*). However, measurement of the chemical profiles could not ensure whether the $CO_2$ and $S^{2-}$ production is due to AOM or not. Therefore, other complementary methods such as *in vitro* incubation with stable isotopes or radioisotopes (e.g. $^{13}CH_4$ and $^{12}CH_4$) and profile measurement of labeled carbon are used for the estimation of AOM rate (by monitoring the $^{13}CH_4$ and $^{13}CO_2$ production in a batch) (Knittel & Boetius, 2009; Reeburgh, 2007). In addition, identification of the microbial community ensures the presence of ANME and establishes the link between the identity of the microorganisms and the AOM activity. A wide range of AOM rates have been observed in the different ANME habitats and bioreactors enrichments (Tables 2.1, 2.2 and 2.3).

### 2.6.2 ANME identification

Specific lipid biomarkers and stable carbon isotopes are often measured at potential AOM sites since the discovery of AOM (Blumenberg et al., 2005; Blumenberg et al., 2004; Hinrichs & Boetius, 2003; Hinrichs et al., 2000; Pancost et al., 2001; Rossel et al., 2008). Biomarkers are used to differentiate between archaeal and bacterial cells. Phospholipids fatty acids with an ether linkage are usually common for bacteria and eukarya (Niemann & Elvert, 2008). Distinction between ANME-1, ANME-2 and ANME-3 was explored by analysis of non-polar lipids and intact polar lipids as biomarkers (Rossel et al., 2008). ANME-1 contains a majority of isoprenoidal glycerol dialkyl glycerol tetraethers on its lipid profile, whereas ANME-2 and ANME-3 mostly contain phosphate-based polar derivatives of archaeol and hydroxyarchaeol (Niemann & Elvert, 2008; Rossel et al., 2008). While detection of lipid biomarkers provide information on the microorganism's identity, the carbon isotopic composition of the biomarkers provides information on the carbon source and/or metabolic fixation pathway of microbes (Hinrichs & Boetius, 2003).

$CH_4$ in marine environments is generally depleted in $^{13}C$ (carbon stable isotope composition, $\delta^{13}C$, of -50 to -110‰), while $CO_2$ is usually isotopically heavier than $CH_4$. Therefore, $CH_4$ oxidation would result in products which are depleted in $^{13}C$. The finding of highly $^{13}C$ depleted lipids in archaeal biomass ($\delta^{13}C < -60‰$) has been used as indicator of $CH_4$ oxidation (AOM) with concomitant assimilation of carbon derived from the light $CH_4$ ($^{12}C$) by the ANME (Emerson & Hedges, 2008; Hinrichs & Boetius, 2003; Hinrichs et al., 1999; Hinrichs et al., 2000; Martens et al., 1999; Thomsen et al., 2001).

However, the carbon isotopic composition in many AOM habitats is complex. For instance, in cold seeps and vent sediments both $CH_4$ and $CO_2$ are isotopically light. The isotopically light $CO_2$ is produced by chemoautotrophic microbes (Alperin & Hoehler, 2009), while the light $CH_4$ could be due to both methanogenesis and AOM (Pohlman et al., 2008). On the basis of these recent findings, the light isotope composition of the lipids in archaeal biomass indicated that the light carbon content can be assimilated via several processes (Figure 2.10): i) the assimilation from isotopically light $CO_2$ ($^{13}C$ depleted $CH_4$ production rather than oxidation) by ANME, i.e. involvement of ANME-1 and ANME-2 in methanogenesis (Bertram et al., 2013; House et al., 2009), ii) the oxidation of $CH_4$ and utilization of inorganic carbon by ANME, i.e. autotrophic AOM by ANME-1 and ANME-2 (Kellermann et al., 2012), iii) AOM by the assimilation of $^{13}C$ depleted $CH_4$ in ANME, i.e. a common AOM process (Emerson & Hedges, 2008; Hinrichs & Boetius, 2003; Hinrichs et al., 1999; Hinrichs et al., 2000; Martens et al., 1999; Thomsen et al., 2001) and iv) $^{13}C$ depleted $CO_2$ assimilation by methanogens (Vigneron et al., 2015). Therefore, the conventional assumption of $^{13}C$ depleted archaeal biomarkers as a proxy for AOM has to be considered carefully. Moreover, the approach is not always straightforward for the depiction of AOM as light carbon in lipids can also originate from archaeal $CH_4$ production and not only from $CH_4$ oxidation (Alperin & Hoehler, 2009; Londry et al., 2008). Thus, the application of multiple approaches is advantageous for explicit understanding of AOM and ANME.

The confusion due to light lipid biomarkers from multiple carbon metabolisms can be partly overcome if stable isotope probing (SIP) is performed. $^{13}C$ enriched $CH_4$ and $CO_2$ can be used as substrates for *in vitro* incubations with the desired inoculum. Isotopic probing followed by lipids biomarker analysis (lipid-SIP) can be used to identify the carbon assimilation pathways for the microbes under investigation (Kellermann et al., 2012). Autotrophic and heterotrophic carbon assimilation together with lipid formation rates can be determined by dual lipid-SIP, which consists of simultaneous addition of deuterated water and $^{13}C$-labeled inorganic carbon (Wegener et al., 2012). Moreover, the visualization of ANME cells or other molecular detection of ANME can be performed for clear elucidation on ANME occurrence.

Phylogenic analysis of 16s rRNA and *mcr*A genes from marine environments is generally performed for assigning identity to ANME types (Alain et al., 2006; Boetius et al., 2000; Harrison et al., 2009; Knittel et al., 2005; Losekann et al., 2007) (details described in section 2.2.2). ANME cells are quantified by Q-PCR to assess ANME growth in enrichments and DNA finger print for comparison of ANME types among AOM sites (Girguis et al., 2005; Lloyd et al., 2011; Timmers et al., 2015; Wankel et al., 2012a). In the recent past, ANME specific primers were designed to enhance the quantification of particular ANME types (Miyashita et al., 2009; Zhou et al., 2014). Q-PCR can be performed to quantify the RNA fraction of ANME genes, thus basically quantifying the active cells (Lloyd et al., 2010). Moreover, quantification of key functional genes such as *mcr*A (methanogenesis related) genes in ANME (Lee et al., 2013; Yanagawa et al., 2011) and *dsr*A (SR related) genes in SRB (Lee et al., 2013) were performed in recent studies. The analysis and quantification of specific functional genes allows the quantification of the microbes expressing the specific function only, so it will be easier to interpret the quantification results. The gene based analysis can nevertheless sometimes leads to a false conclusion. For example, the findings of specific DNA/RNA in a certain location may not always indicate the active cells in that location because cells might be transported from nearby active AOM areas. Hence, activity measurements of the biomass in those locations over time remain essential.

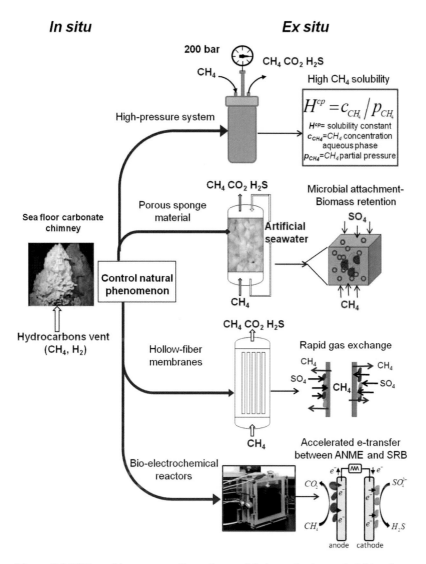

**Figure 2.9** Different bioreactor configurations and their mechanisms mimicking the growth mode of ANME in natural habitats to enhance *ex situ* growth of ANME.

Recent studies on AOM pursued high throughput shot gun sequencing for the analysis of archaeal and bacterial communities in different sites including high temperature AOM (Mason et al., 2015; Wankel et al., 2012a). Small subunit ribosomal RNA (SSU rRNA) genes containing highly conserved and variable regions (V1-V9) were used as a marker for the high through put sequencing (Lynch & Neufeld, 2015). Among 23 distinct $CH_4$ seep sediments studied via pyrotag library analysis, ANME archaea and Seep-SRB bacteria appeared as major communities in the cold anaerobic $CH_4$ seep, whereas aerobic methanotrophs and sulfide oxidizing *Thiotricales* groups were found mostly in the oxic part of $CH_4$ seeps (Ruff et al., 2015).

High throughput sequencing of specific gene amplicons provides information about the microbial community composition, whereas whole genomics analysis explores the functional profiles from gene to family level. Thus, the community metabolic pathways can be constructed on the basis of these genes (Franzosa et al., 2015). Chistoserdova *et al.* (Chistoserdova, 2015) reviewed the aerobic and anaerobic methanotrophy on the basis of metagenomic studies. Metagenomics of ANME-1 and ANME-2a have been performed so far, supporting the reverse methanogenesis pathway for $SO_4^{2-}$ dependent AOM (Hallam et al., 2004; Meyerdierks et al., 2010). Moreover, the nitrate dependent AOM pathway was depicted by the ANME-2d genome in which the nitrate reductase specific gene was highly expressed (Haroon et al., 2013). Yet, more details on the omics based analysis of other ANME-phylotypes should be explored. Nevertheless, complete genomic studies are relevant for highly enriched ANME communities rather than the genomic analysis with sediment containing a few ANME cells for the explicit interpretation of genomic data and metabolic pathways. It should be noted that the genomic data provide mostly the information of the dominant community, so it is difficult to extract the information from the ANME genome if the amount of ANME genes is low in the sample analyzed.

### 2.6.3   ANME visualization and their functions studies

FISH images provide insights regarding the morphology and the spatial arrangement of ANME and their bacterial associates within aggregates (Blumenberg et al., 2004; Boetius et al., 2000; Knittel et al., 2005; Roalkvam et al., 2011). The FISH method has been widely discussed and applied in the past 25 years (Amann & Fuchs, 2008). Catalyzed reported deposition–fluorescence *in situ* hybridization (CARD-FISH) with horseradish peroxidase (HRP) labeled probes are commonly used to visualize the ANME cells and SRBs in marine sediments (Holler et al., 2011b; Lloyd et al., 2011). The signal is amplified in CARD-FISH by using these probes together with fluorescently labeled tyramides, therefore copious fluorescent molecules can be introduced and the sensitivity increases compared to normal FISH (Pernthaler Annelie 2002). Detection of a few ANME cells by FISH does not always means that the studied site is an ANME habitat. For explicit AOM illustration, it is essential to combine FISH with other approaches such as activity measurements, quantification of ANME cells (by cell count, Q-PCR or quantification FISH), isotope probing or spectroscopic detection of metabolites.

In order to link the identity of microorganism to their functions, other methods that investigate the physiology and activities have to be combined with FISH. FISH in combination with SIMS (secondary ion mass spectroscopy) was used in AOM and ANME studies (Orphan et al., 2001b), which provides the linkage of ANME to its function by the visualization of ANME and analysis of compounds assimilated in the cells. The SIMS can analyze the isotopic composition of the cells so that it can be used to understand the mechanisms of AOM along with the syntrophy and intermediates (Orphan & House, 2009).

NanoSIMS (Nanometer-scale secondary ion mass spectrometry) miniaturized SIMS instrumentation with a sub micrometer spatial resolution has been used for ANME studies. It allows observation of single cell morphology in combination with FISH and quantitative analysis of the elemental and isotopic composition of cells with high sensitivity and precision

(Behrens et al., 2008; Musat et al., 2008; Polerecky et al., 2012). FISH-NanoSIMS has been used in ANME studies detailing nitrogen fixation by ANME-2d archaea (Dekas et al., 2009) and sulfur metabolism in ANME-2 cells (Milucka et al., 2012). Normally highly enriched microbial communities are incubated with isotopic labeled substrates and the fate of the substrates is detected by specifically designed NanoSIMS equipment. FISH-NanoSIMS is often complemented by advanced microscopic observations such as scanning (SEM) or transmission (TEM) electron microscopy or atomic force microscopy (AFM) (Polerecky et al., 2012). Recently it was described how combining FISH-NanoSIMS with SIP can be used to link identity, function and metabolic activity at cellular level and therefore showing the metabolic interactions within consortia (Musat et al., 2016).

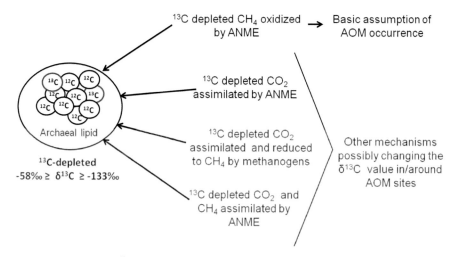

**Figure 2.10** Detection of $^{13}C$ depleted lipids in Archaea as proxy of $CH_4$ oxidation: Basic assumption of AOM occurrence and other possible mechanisms that induce a change in the $\delta^{13}C$ value in archaeal lipids.

Microautoradiography-FISH (MAR-FISH) is a promising approach to study ANME physiology by monitoring the assimilation of radio-labeled substrates by individual cells. The radio-labeled substrates (e.g. different carbon sources) are added to the samples containing active microbes and the fate of radioisotopes can be detected by MAR with simultaneous microbial identification by FISH (Lee et al., 1999; Nielsen & Nielsen, 2010). The handling of radioisotopes can limit the applications of this powerful technique.

Another appealing method for AOM studies is Raman-FISH (Wagner, 2009), which analyzes the stable isotope at a micrometer level to provide the ecophysiology of a single cell. Raman microspectroscopy detects and quantifies the stable or radio isotope labeled substrate assimilation in the cell under study. The Raman spectra wavelength can provide distinction between the uptake of different substrates among cells exhibiting different metabolic pathways (Wagner, 2009). Raman spectroscopy can detect the molecular composition of a cell and thus can provide information on the molecules which are assimilated in the cells. The technique is thus useful for the study of assimilation of carbon and sulfur compounds in ANME. ANME cells can be visualized by FISH, and then examined with Raman spectroscopy for the substrate assimilation up to single cell level. It is highly applicable for the study of ANME cells as single cells from a complex microbial consortium can be analyzed and the assimilated compounds by these cells can be monitored.

### 2.6.4 New study approaches to AOM

ANME studies have immensely benefited from the advancement of microscopic and molecular tools. In recent years, several complementary approaches were applied for depicting AOM mechanisms, such as FISH-NanoSIMS together with Raman-FISH in a highly enriched ANME community (Milucka et al., 2012; Musat et al., 2016) and metagenomics together with FISH and continuous enrichment activity assays (Haroon et al., 2013). There are still several open questions to be addressed in ANME studies such as identification of intermediates, alternative substrates, tolerance limit for various environmental stresses and exploration of several ANME habitats. Also details of the carbon and sulfur metabolism by ANME and SRB are not elucidated till date.

Despite of advancement in genomic sequencing, the genome of only some phylotypes of ANME (ANME-2a,-2d and ANME-1) has been studied and the metabolic pathways were predicted. The predicted metabolic pathways by genomics can be verified by ecophysiological studies in combination with SIMS based spectroscopy. Further, many prospective approaches could be used for ANME studies to explore the ANME mechanism and ecophysiology, for example, an *in situ* SIP based survey for the study of AOM occurrence and carbon assimilation, single cell genomics for predicting metabolic pathways and genes from single cell isolates (Rinke et al., 2014), imaging and mass spectroscopy of single cells or aggregates for understanding the metabolisms (Watrous & Dorrestein, 2011), and atomic force microscopy for the study of ANME cellular structure and detection of the effect of different stresses on the cell membrane (Dufrêne, 2014). Note that all these studies are only possible if the ANME microbial mats are appropriately handled from the seafloor to the laboratory or enriched in bioreactors.

In view of the complexity and lengthiness to cultivate a sufficient amount of ANME biomass, *in situ* investigations with sophisticated *in situ* laboratory might overcome the current biomass handling and enrichment limitations. Taking advantage on the latest advances in deep-sea instrumentation, which include various on-line data acquisition instruments, it should be possible for example to conduct *in situ* gas push-pull tests (i.e., tracer tests) (Urmann et al., 2004) which combined with *in situ* stable isotope probing approaches (Wankel et al., 2012b) and *in situ* molecular analysis can yield detailed microbial activity and function measurements in tandem with microbial identity. A deep-sea environmental sample processor can be stationed from near surface ocean water to 1000 m depth. It is capable to detect *in situ* and in real time molecular signals indicative of certain microorganisms or genes (Paul et al., 2007; Scholin et al., 2009). As proof-of-concept, an environmental sample processor has been used for the quantitative detection of 16S rRNA and particulate $CH_4$ monooxygenase (*pmo*A) genes of aerobic methanotrophs near a $CH_4$-rich mound at a water depth of about 800 m (Ussler et al., 2013). In principle, the environmental sample processor can be configured to detect and quantify genes and gene products from a wide range of microbial types (Preston et al., 2011). Additionally, the environmental sample processor is able to store samples for later *ex situ* validation analysis. The long term deployment capacity of the environmental sample processor is under development and this capability should allow temporal profiling of microorganisms which in tandem with on line characterization of physico-chemical parameters may help to understand which drivers are most important for the proliferation of active ANME communities in deep-sea.

## 2.7    Conclusion and outlook

Undoubtedly much was learned about AOM in the last four decades, yet key knowledge gaps still exist. One of the most remarkable aspects requiring investigation relates to the proposed syntrophic association between ANME and SRB. Overall whether, when and how AOM occurs in obligatory syntrophic association with SRB remains unclear. If such syntrophy occurs through direct electron transfer as proposed by McGlynn et al. (2015) and Wegener et al. (2015), it is necessary to understand the role of the cytochromes in ANME metabolism and the function of the pili-like structures observed. It has been shown that AOM does not necessarily occur in a syntrophic association, ANME can be decoupled from the bacterial partner in the laboratory (Scheller et al., 2016), showing the possibility to grow ANME seperately and understand its metabolism.

The marine habitats hosting ANME have been widely explored in the past, details on niche differentiation among the various ANME clades need to be further assessed. The presence and relevance of $SO_4^{2-}$ dependent AOM in freshwater environments requires further exploration. Although a few investigations on freshwater habitats have been conducted, unambiguous links between the presence and activity of AOM are still required. Sediments from eutrophic lakes and freshwater tidal creeks might be suitable locations to explore (Sivan et al., 2011). Yet, another aspect to resolve is the existence and identity of ANME directly utilizing iron or manganese oxides as electron acceptors. Ettwig et al. (2016) demonstrates that iron and manganese dependent $CH_4$ oxidation occurred in freshwater ecosystems. Scheller et al. (2016) showed that marine samples containing ANME-2 could couple the reduction of chelated oxidized iron, but whether ANME can use also metal oxides found in marine sediment, it still need to be proved. Some marine and brackish coastal locations having abundant iron oxides within $CH_4$ rich sediments have been identified. Sediments from those locations appear suitable for harboring ANME (Egger et al., 2015; Riedinger et al., 2014; Wankel et al., 2012a). Moreover, other naturally occurring electron acceptors such as selenate can be investigated in future AOM studies.

After a long effort, the most incommoding drawback is not being able to readily obtain enrichments of $SO_4^{2-}$ dependent ANME. With the exception of a few studies in which ANME-2a enrichments were obtained after eight (Milucka et al., 2012) and three years (Meulepas et al., 2009a) in bioreactors, most biochemical studies have been conducted using naturally ANME enriched sediments of which the retrieved small quantities often limit experimental tests. In such context, proper handling of ANME biomass from the seafloor to the laboratory as well as the enrichment in bioreactor configurations mimicking *in situ* conditions are in priority. Alternatively, single cell microscopy and genomics as well as the development of an advanced of *in situ* deep-sea laboratory can help in unrevealing some of the remaining unknowns of the ANME metabolisms and ecophysiology.

## 2.8 References

Alain, K., Holler, T., Musat, F., Elvert, M., Treude, T., Krüger, M. 2006. Microbiological investigation of methane- and hydrocarbon-discharging mud volcanoes in the Carpathian Mountains, Romania. *Environ. Microbiol.*, **8**(4), 574-590.

Aller, R.C., Rude, P. 1988. Complete oxidation of solid phase sulfides by manganese and bacteria in anoxic marine sediments. *Geochim. Cosmochim. Ac.*, **52**(3), 751-765.

Alperin, M., Hoehler, T. 2010. The ongoing mystery of sea-floor methane. *Science*, **329**(5989), 288-289.

Alperin, M.J., Hoehler, T.M. 2009. Anaerobic methane oxidation by archaea/sulfate-reducing bacteria aggregates: 2. Isotopic constraints. *Am. J. Sci.*, **309**(10), 958-984.

Amann, R., Fuchs, B.M. 2008. Single-cell identification in microbial communities by improved fluorescence *in situ* hybridization techniques. *Nature Rev. Microbiol.*, **6**(5), 339-348.

Amann, R.I., Ludwig, W., Schleifer, K.H. 1995. Phylogenetic identification and *in situ* detection of individual microbial cells without cultivation. *Microbiol. Rev.*, **59**(1), 143-169.

Aoki, M., Ehara, M., Saito, Y., Yoshioka, H., Miyazaki, M., Saito, Y., Miyashita, A., Kawakami, S., Yamaguchi, T., Ohashi, A., Nunoura, T., Takai, K., Imachi, H. 2014. A long-term cultivation of an anaerobic methane-oxidizing microbial community from deep-sea methane-seep sediment using a continuous-flow bioreactor. *PLoS ONE*, **9**(8), Pe105356.

Archer, D., Buffett, B., Brovkin, V. 2009. Ocean methane hydrates as a slow tipping point in the global carbon cycle. *Proc. Natl. Acad. Sci. USA*, **106**(49), 20596-20601.

Beal, E.J., House, C.H., Orphan, V.J. 2009. Manganese- and iron-dependent marine methane oxidation. *Science*, **325**(5937), 184-187.

Behrens, S., Lösekann, T., Pett-Ridge, J., Weber, P.K., Ng, W.-O., Stevenson, B.S., Hutcheon, I.D., Relman, D.a., Spormann, A.M. 2008. Linking microbial phylogeny to metabolic activity at the single-cell level by using enhanced element labeling-catalyzed reporter deposition fluorescence *in situ* hybridization (EL-FISH) and NanoSIMS. *Appl. Environ. Microbiol.*, **74**(10), 3143-3150.

Bertram, S., Blumenberg, M., Michaelis, W., Siegert, M., Krüger, M., Seifert, R. 2013. Methanogenic capabilities of ANME-archaea deduced from [13]C-labelling approaches. *Environ. Microbiol.*, **15**(8), 2384-2393.

Biddle, J.F., Cardman, Z., Mendlovitz, H., Albert, D.B., Lloyd, K.G., Boetius, A., Teske, A. 2012. Anaerobic oxidation of methane at different temperature regimes in Guaymas Basin hydrothermal sediments. *ISME J.*, **6**(5), 1018-1031.

Blazewicz, S.J., Petersen, D.G., Waldrop, M.P., Firestone, M.K. 2012. Anaerobic oxidation of methane in tropical and boreal soils: Ecological significance in terrestrial methane cycling. *J. Geophys. Res. Biogeosci.*, **117**(G2), 1-9.

Blumenberg, M., Seifert, R., Nauhaus, K., Pape, T., Michaelis, W. 2005. *In vitro* study of lipid biosynthesis in an anaerobically methane-oxidizing microbial mat. *Appl. Environ. Microbiol.*, **71**(8), 4345-4351.

Blumenberg, M., Seifert, R., Reitner, J., Pape, T., Michaelis, W. 2004. Membrane lipid patterns typify distinct anaerobic methanotrophic consortia. *Proc. Natl. Acad. Sci. USA*, **101**(30), 11111-11116.

Boetius, A., Ravenschlag, K., Schubert, C.J., Rickert, D., Widdel, F., Gieseke, A., Amann, R., Jørgensen, B.B., Witte, U., Pfannkuche, O. 2000. A marine microbial consortium apparently mediating anaerobic oxidation of methane. *Nature*, **407**(6804), 623-626.

Boetius, A., Wenzhöfer, F. 2013. Seafloor oxygen consumption fuelled by methane from cold seeps. *Nat. Geosc.*, **6**(9), 725-734.

Bradley, A.S., Fredricks, H., Hinrichs, K.-U., Summons, R.E. 2009. Structural diversity of diether lipids in carbonate chimneys at the Lost City hydrothermal field. *Org. Geochem.*, **40**(12), 1169-1178.

Brazelton, W.J., Schrenk, M.O., Kelley, D.S., Baross, J.A. 2006. Methane- and sulfur-metabolizing microbial communities dominate the Lost City hydrothermal field ecosystem. *Appl. Environ. Microbiol.*, **72**(9), 6257-6270.

Brooks, J.M., Field, M.E., Kennicutt, M.C. 1991. Observations of gas hydrates in marine sediments, offshore northern California. *Mar. Geol.*, **96**(1), 103-109.

Buffett, B., Archer, D. 2004. Global inventory of methane clathrate: sensitivity to changes in the deep ocean. *Earth Planet. Sci. Lett.*, **227**(3-4), 185-199.

Canfield, D.E., Thamdrup, B., Hansen, J.W. 1993. The anaerobic degradation of organic matter in Danish coastal sediments: iron reduction, manganese reduction, and sulfate reduction. *Geochim. Cosmochim. Ac.*, **57**(16), 3867-3683.

Cassidy, J., Lubberding, H.J., Esposito, G., Keesman, K.J., Lens, P.N.L. 2015. Automated biological sulphate reduction: a review on mathematical models, monitoring and bioprocess control. *FEMS Microbiol. Rev.*, **39**(6), 823-853.

Chistoserdova, L. 2015. Methylotrophs in natural habitats: current insights through metagenomics. *Appl. Environ. Microbiol.*, **99**(14), 5763-5779.

Chistoserdova, L., Vorholt, J.A., Lidstrom, M.E. 2005. A genomic view of methane oxidation by aerobic bacteria and anaerobic archaea. *Genome Biol.*, **6**(2), 5763-5779.

D'Hondt, S., Jørgensen, B.B., Miller, D.J., Batzke, A., Blake, R., Cragg, B.A., Cypionka, H., Dickens, G.R., Ferdelman, T., Hinrichs, K.-U., Holm, N.G., Mitterer, R., Spivack, A., Wang, G., Bekins, B., Engelen, B., Ford, K., Gettemy, G., Rutherford, S.D., Sass, H., Skilbeck, C.G., Aiello, I.W., Guèrin, G., House, C.H., Inagaki, F., Meister, P., Naehr, T., Niitsuma, S., Parkes, R.J., Schippers, A., Smith, D.C., Teske, A., Wiegel, J., Padilla, C.N., Acosta, J.L.S. 2004. Distributions of microbial activities in deep subseafloor sediments. *Science*, **306**(5705), 2216-2221.

D'Hondt, S., Rutherford, S., Spivack, A.J. 2002. Metabolic activity of subsurface life in deep-sea sediments. *Science*, **295**(5562), 2067-2070.

Dahl, C., Prange, A. 2006. Bacterial sulfur globules: occurrence, structure and metabolism. in: J. Shively (Ed.), *Inclusions in Prokaryotes*, Vol. 1, Springer Berlin Heidelberg, Germany, pp. 21-51.

Dale, A.W., Van Cappellen, P., Aguilera, D.R., Regnier, P. 2008. Methane efflux from marine sediments in passive and active margins: estimations from bioenergetic reaction-transport simulations. *Earth Planet. Sci. Lett.*, **265**(3-4), 329-344.

Dekas, A.E., Chadwick, G.L., Bowles, M.W., Joye, S.B., Orphan, V.J. 2014. Spatial distribution of nitrogen fixation in methane seep sediment and the role of the ANME archaea. *Environ. Microbiol.*, **16**(10), 3012-3029.

Dekas, A.E., Poretsky, R.S., Orphan, V.J. 2009. Deep-Sea archaea fix and share nitrogen in methane-consuming microbial consortia. *Science*, **326**(5951), 422-426.

Deusner, C., Meyer, V., Ferdelman, T. 2009. High-pressure systems for gas-phase free continuous incubation of enriched marine microbial communities performing anaerobic oxidation of methane. *Biotechnol. Bioeng.*, **105**(3), 524-533.

Dufrêne, Y.F. 2014. Atomic force microscopy in microbiology: new structural and functional insights into the microbial cell surface. *MBio*, **5**(4), e01363.

Durisch-Kaiser, E., Klauser, L., Wehrli, B., Schubert, C. 2005. Evidence of intense archaeal and bacterial methanotrophic activity in the Black Sea water column. *Appl. Environ. Microbiol.*, **71**(12), 8099-8106.

Egger, M., Rasigraf, O., Sapart, C.l.J., Jilbert, T., Jetten, M.S., Röckmann, T., van der Veen, C., Banda, N., Kartal, B., Ettwig, K.F., Slomp, C.P. 2015. Iron-mediated anaerobic oxidation of methane in brackish coastal sediments. *Environ. Sci. Technol.*, **49**(1), 277-283.

Emerson, S., Hedges, J. 2008. Stable and radioactive isotopes. in: *Chemical oceanography and the marine carbon cycle*, (1st edition), Cambridge University Press, pp. 468.

Ettwig, K.F., Butler, M.K., Le Paslier, D., Pelletier, E., Mangenot, S., Kuypers, M.M.M., Schreiber, F., Dutilh, B.E., Zedelius, J., de Beer, D., Gloerich, J., Wessels, H.J.C.T., van Alen, T., Luesken, F., Wu, M.L., van de Pas-Schoonen, K.T., Op den Camp, H.J.M., Janssen-Megens, E.M., Francoijs, K.-J., Stunnenberg, H., Weissenbach, J., Jetten, M.S.M., Strous, M. 2010. Nitrite-driven anaerobic methane oxidation by oxygenic bacteria. *Nature*, **464**(7288), 543-548.

Ettwig, K.F., Zhu, B., Speth, D., Keltjens, J.T., Jetten, M.S.M., Kartal, B. 2016. Archaea catalyze iron-independent anaerobic oxidation of methane. *Proc. Natl. Acad. Sci. USA*, **13**(15), 12792-12796.

Finster, K. 2008. Microbiological disproportionation of inorganic sulfur compounds. *J. Sulfur Chem.*, **29**(3-4), 281-292.

Franzosa, E.A., Hsu, T., Sirota-Madi, A., Shafquat, A., Abu-Ali, G., Morgan, X.C., Huttenhower, C. 2015. Sequencing and beyond: integrating molecular 'omics' for microbial community profiling. *Nat. Rev. Microbiol.*, **13**(6), 360-372.

Gauthier, M., Bradley, R.L., Šimek, M. 2015. More evidence that anaerobic oxidation of methane is prevalent in soils: is it time to upgrade our biogeochemical models? *Soil Biol. Biochem.*, **80**(9), 167-174.

Girguis, P.R., Cozen, A.E., DeLong, E.F. 2005. Growth and population dynamics of anaerobic methane-oxidizing archaea and sulfate-reducing bacteria in a continuous-flow bioreactor. *Appl. Environ. Microbiol.*, **71**(7), 3725-3733.

Girguis, P.R., Orphan, V.J., Hallam, S.J., DeLong, E.F. 2003. Growth and methane oxidation rates of anaerobic methanotrophic archaea in a continuous-flow bioreactor. *Appl. Environ. Microbiol.*, **69**(9), 5472-5482.

Gonzalez-Gil, G., Meulepas, R.J.W., Lens, P.N.L. 2011. Biotechnological aspects of the use of methane as electron donor for sulfate reduction. in: Murray, M.-Y. (Ed.), *Comprehensive Biotechnology*, Vol. 6 (2nd edition), Elsevier B.V. Amsterdam, the Netherlands, pp. 419-434.

Hallam, S.J., Girguis, P.R., Preston, C.M., Richardson, P.M., DeLong, E.F. 2003. Identification of methyl coenzyme M reductase A (mcrA) genes associated with methane-oxidizing archaea. *Appl. Environ. Microbiol.*, **69**(9), 5483-5491.

Hallam, S.J., Putnam, N., Preston, C.M., Detter, J.C., Rokhsar, D., Richardson, P.M., DeLong, E.F. 2004. Reverse methanogenesis: testing the hypothesis with environmental genomics. *Science*, **305**(5689), 1457-1462.

Hanson, R.S., Hanson, T.E. 1996. Methanotrophic bacteria. *Microbiol. Rev.*, **60**(2), 439-471.

Harder, J. 1997. Anaerobic methane oxidation by bacteria employing [14]C-methane uncontaminated with [14]C-carbon monoxide. *Mar. Geol.*, **137**(1-2), 13-23.

Haroon, M.F., Hu, S., Shi, Y., Imelfort, M., Keller, J., Hugenholtz, P., Yuan, Z., Tyson, G.W. 2013. Anaerobic oxidation of methane coupled to nitrate reduction in a novel archaeal lineage. *Nature*, **500**(7468), 567-570.

Harrison, B.K., Zhang, H., Berelson, W., Orphan, V.J. 2009. Variations in archaeal and bacterial diversity associated with the sulfate-methane transition zone in continental margin sediments (Santa Barbara Basin, California). *Appl. Environ. Microbiol.*, **75**(6), 1487-1499.

Haynes, C.A., Gonzalez, R. 2014. Rethinking biological activation of methane and conversion to liquid fuels. *Nat. Chem. Biol.*, **10**(5), 331-339.

Heijs, S.K., Haese, R.R., van der Wielen, P.W., Forney, L.J., van Elsas, J.D. 2007. Use of 16S rRNA gene based clone libraries to assess microbial communities potentially involved in anaerobic methane oxidation in a Mediterranean cold seep. *Microb. Ecol.*, **53**(3), 384-398.

Hinrichs, K.-U., Boetius, A. 2003. The anaerobic oxidation of methane: new insights in microbial ecology and biogeochemistry. in: Wefer, G., Billett, D., Hebbeln, D., Jørgensen, B.B., Schlüter, M., van Weering, T.E. (Eds), *Ocean Margin Systems*, Springer Berlin Heidelberg, Germany, pp. 457-477.

Hinrichs, K.-U., Hayes, J.M., Sylva, S.P., Brewer, P.G., DeLong, E.F. 1999. Methane-consuming archaebacteria in marine sediments. *Nature*, **398**(6730), 802-805.

Hinrichs, K.-U., Summons, R.E., Orphan, V.J., Sylva, S.P., Hayes, J.M. 2000. Molecular and isotopic analysis of anaerobic methane-oxidizing communities in marine sediments. *Org. Geochem.*, **31**(12), 1685-1701.

Hoehler, T.M., Alperin, M.J., Albert, D.B., Martens, S. 1994. Field and laboratory studies of methane oxidation in an anoxic marine sediment: evidence for a methanogen-sulfate reducer consortium. *Global. Biogeochem. Cy.*, **8**(4), 451-463.

Holler, T., Wegener, G., Niemann, H., Deusner, C., Ferdelman, T.G., Boetius, A., Brunner, B., Widdel, F. 2011a. Carbon and sulfur back flux during anaerobic microbial oxidation of methane and coupled sulfate reduction. *Proc. Natl. Acad. Sci. USA*, **108**(52), E1484-E1490.

Holler, T., Widdel, F., Knittel, K., Amann, R., Kellermann, M.Y., Hinrichs, K.-U., Teske, A., Boetius, A., Wegener, G. 2011b. Thermophilic anaerobic oxidation of methane by marine microbial consortia. *ISME J.*, **5**(12), 1946-1956.

House, C.H., Orphan, V.J., Turk, K., A., Thomas, B., Pernthaler, A., Vrentas, J.M., Joye, S., B. . 2009. Extensive carbon isotopic heterogeneity among methane seep microbiota. *Environ. Microbiol.*, **11**(9), 2207-2215.

Hu, S., Zeng, R.J., Burow, L.C., Lant, P., Keller, J., Yuan, Z. 2009. Enrichment of denitrifying anaerobic methane oxidizing microorganisms. *Environ. Microbiol. Rep.*, **1**(5), 377-384.

Imachi, H., Aoi, K., Tasumi, E., Saito, Y.Y., Yamanaka, Y., Yamaguchi, T., Tomaru, H., Takeuchi, R., Morono, Y., Inagaki, F., Takai, K. 2011. Cultivation of methanogenic community from subseafloor sediments using a continuous-flow bioreactor. *ISME J.*, **5**(12), 1913-1925.

IPCC. 2007. Climate change 2007: the physical science basis. in: Solomon, S., Qin, D., Manning, M., Chen, Z., Marquis, M., Averyt, K.B., Tignor, M., Miller, H.L. (Eds.), *Contribution of working group I to the fourth assessment report of the intergovernmental panel on climate change*, Cambridge University Press, New York, USA., pp. 996.

Iversen, N., Jørgensen, B.B. 1985. Anaerobic methane oxidation rates at the sulfate-methane transition in marine sediments from Kattegat and Skagerrak (Denmark). *Limnol. Oceanogr.*, **30**(5), 944-955.

Jagersma, G.C., Meulepas, R.J.W., Heikamp-de Jong, I., Gieteling, J., Klimiuk, A., Schouten, S., Sinninghe Damsté, J.S., Lens, P.N.L., Stams, A.J. 2009. Microbial diversity and community structure of a highly active anaerobic methane-oxidizing sulfate-reducing enrichment. *Environ. Microbiol.*, **11**(12), 3223-3232.

Jørgensen, B.B., Kasten, S. 2006. Sulfur cycling and methane oxidation. in: Schulz, H., Zabel, M. (Eds.), *Marine Geochemistry*, (2$^{nd}$ edition), Springer Berlin Heidelberg, Germany, pp. 271-309.

Joye, S.B., Boetius, A., Orcutt, B.N., Montoya, J.P., Schulz, H.N., Erickson, M.J., Lugo, S.K. 2004. The anaerobic oxidation of methane and sulfate reduction in sediments from Gulf of Mexico cold seeps. *Chem. Geol.*, **205**(3-4), 219-238.

Kato, S., Hashimoto, K., Watanabe, K. 2012. Microbial interspecies electron transfer via electric currents through conductive minerals. *Proc. Natl. Acad. Sci. USA*, **109**(25), 10042-10046.

Kellermann, M.Y., Wegener, G., Elvert, M., Yoshinaga, M.Y., Lin, Y.-S., Holler, T., Mollar, X.P., Knittel, K., Hinrichs, K.-U. 2012. Autotrophy as a predominant mode of carbon fixation in anaerobic methane-oxidizing microbial communities. *Proc. Natl. Acad. Sci. USA,* **109**(47), 19321-19326.

Kirschke, S., Bousquet, P., Ciais, P., Saunois, M., Canadell, J.G., Dlugokencky, E.J., Bergamaschi, P., Bergmann, D., Blake, D.R., Bruhwiler, L., Cameron-Smith, P., Castaldi, S., Chevallier, F., Feng, L., Fraser, A., Heimann, M., Hodson, E.L., Houweling, S., Josse, B., Fraser, P.J., Krummel, P.B., Lamarque, J.-F., Langenfelds, R.L., Le Quere, C., Naik, V., O'Doherty, S., Palmer, P.I., Pison, I., Plummer, D., Poulter, B., Prinn, R.G., Rigby, M., Ringeval, B., Santini, M., Schmidt, M., Shindell, D.T., Simpson, I.J., Spahni, R., Steele, L.P., Strode, S.A., Sudo, K., Szopa, S., van der Werf, G.R., Voulgarakis, A., van Weele, M., Weiss, R.F., Williams, J.E., Zeng, G. 2013. Three decades of global methane sources and sinks. *Nat. Geosci.*, **6**(10), 813-823.

Knittel, K., Boetius, A. 2009. Anaerobic oxidation of methane: progress with an unknown process. *Annu. Rev. Microbiol.*, **63**(1), 311-334.

Knittel, K., Lösekann, T., Boetius, A., Kort, R., Amann, R. 2005. Diversity and distribution of methanotrophic archaea at cold seeps. *Appl. Environ. Microbiol.*, **71**(1), 467-479.

Kormas, K., Meziti, A., Dählmann, A., De Lange, G., Lykousis, V. 2008. Characterization of methanogenic and prokaryotic assemblages based on mcrA and 16S rRNA gene diversity in sediments of the Kazan mud volcano (Mediterranean Sea). *Geobiology*, **6**(5), 450-460.

Krause, S., Steeb, P., Hensen, C., Liebetrau, V., Dale, A.W., Nuzzo, M., Treude, T. 2014. Microbial activity and carbonate isotope signatures as a tool for identification of spatial differences in methane advection: a case study at the Pacific Costa Rican margin. *Biogeosciences*, **11**(2), 507-523.

Krüger, M., Blumenberg, M., Kasten, S., Wieland, A., Känel, L., Klock, J.-H., Michaelis, W., Seifert, R. 2008. A novel, multi-layered methanotrophic microbial mat system growing on the sediment of the Black Sea. *Environ. Microbiol.*, **10**(8), 1934-1947.

Krüger, M., Wolters, H., Gehre, M., Joye, S.B., Richnow, H.-H. 2008. Tracing the slow growth of anaerobic methane-oxidizing communities by $^{15}$N-labelling techniques. *FEMS Microbiol. Ecol.*, **63**(3), 401-411.

Lanoil, B.D., Sassen, R., La Duc, M.T., Sweet, S.T., Nealson, K.H. 2001. Bacteria and archaea physically associated with Gulf of Mexico gas hydrates. *Appl. Environ. Microbiol.*, **67**(11), 5143-5153.

Larowe, D.E., Dale, A.W., Regnier, P. 2008. A thermodynamic analysis of the anaerobic oxidation of methane in marine sediments. *Geobiology*, **6**(5), 436-449.

Lazar, C.S., John Parkes, R., Cragg, B.A., L'Haridon, S., Toffin, L. 2012. Methanogenic activity and diversity in the centre of the Amsterdam Mud Volcano, Eastern Mediterranean Sea. *FEMS Microbiol. Ecol.*, **81**(1), 243-254.

Lee, J.-W., Kwon, K.K., Azizi, A., Oh, H.-M., Kim, W., Bahk, J.-J., Lee, D.-H., Lee, J.-H. 2013. Microbial community structures of methane hydrate-bearing sediments in the Ulleung Basin, East Sea of Korea. *Mar. Petrol. Geol.*, **47**, 136-146.

Lee, N., Nielsen, P.H., Andreasen, K.H., Juretschko, S., Nielsen, J.L., Schleifer, K.-H., Wagner, M. 1999. Combination of fluorescent *in situ* hybridization and microautoradiography-a new tool for structure-function analyses in microbial ecology. *Appl. Environ.l Microbiol.*, **65**(3), 1289-1297.

Lens, P.N.L., Vallero, M., Esposito, G., Zandvoort, M.H. 2002. Perspectives of sulfate reducing bioreactors in environmental biotechnology. *Rev. Environ. Sci. Biotechnol.*, **1**(4), 311-325.

Lever, M.A., Rouxel, O., Alt, J.C., Shimizu, N., Ono, S., Coggon, R.M., Shanks, W.C., Lapham, L., Elvert, M., Prieto-Mollar, X., Hinrichs, K.-U., Inagaki, F., Teske, A. 2013. Evidence for microbial carbon and sulfur cycling in deeply buried ridge flank basalt. *Science*, **339**(6125), 1305-1308.

Lloyd, K.G., Alperin, M.J., Teske, A. 2011. Environmental evidence for net methane production and oxidation in putative ANaerobic MEthanotrophic (ANME) archaea. *Environ. Microbiol.*, **13**(9), 2548-2564.

Lloyd, K.G., Lapham, L., Teske, A. 2006. An anaerobic methane-oxidizing community of ANME-1b archaea in hypersaline Gulf of Mexico sediments. *Appl. Environ. Microbiol.*, **72**(11), 7218-7230.

Lloyd, K.G., MacGregor, B.J., Teske, A. 2010. Quantitative PCR methods for RNA and DNA in marine sediments: maximizing yield while overcoming inhibition. *FEMS Microbiol. Ecol.*, **72**(1), 143-151.

Londry, K.L., Dawson, K.G., Grover, H.D., Summons, R.E., Bradley, A.S. 2008. Stable carbon isotope fractionation between substrates and products of *Methanosarcina barkeri*. *Org. Geochem.*, **39**(5), 608-621.

Losekann, T., Knittel, K., Nadalig, T., Fuchs, B., Niemann, H., Boetius, A., Amann, R. 2007. Diversity and abundance of aerobic and anaerobic methane oxidizers at the Haakon Mosby mud volcano, Barents Sea. *Appl. Environ. Microbiol.*, **73**(10), 3348-3362.

Lovley, D.R. 2012. Electromicrobiology. *Annu. Rev. Microbiol.*, **66**(1), 391-409.

Lovley, D.R. 2008. Extracellular electron transfer: wires, capacitors, iron lungs, and more. *Geobiology*, **6**(3), 225-231.

Lynch, M.D., Neufeld, J.D. 2015. Ecology and exploration of the rare biosphere. *Nat. Rev. Microbiol.*, **13**(4), 217-229.

Maignien, L., Parkes, R.J., Cragg, B., Niemann, H., Knittel, K., Coulon, S., Akhmetzhanov, A., Boon, N. 2013. Anaerobic oxidation of methane in hypersaline cold seep sediments. *FEMS Microbiol. Ecol.*, **83**(1), 214-231.

Marlow, J.J., Steele, J.A., Case, D.H., Connon, S.A., Levin, L.A., Orphan, V.J. 2014a. Microbial abundance and diversity patterns associated with sediments and carbonates from the methane seep environments of Hydrate Ridge, OR. *Front. Mar. Sci.*, **1**(44), 1-16.

Marlow, J.J., Steele, J.A., Ziebis, W., Thurber, A.R., Levin, L.A., Orphan, V.J. 2014b. Carbonate-hosted methanotrophy represents an unrecognized methane sink in the deep sea. *Nat. Commun.*, **5**(5094), 1-12.

Martens, C.S., Albert, D.B., Alperin, M. 1999. Stable isotope tracing of anaerobic methane oxidation in the gassy sediments of Eckernförde Bay, German Baltic Sea. *A. J. Sci.*, **299**(7-9), 589-610.

Martens, C.S., Berner, R.A. 1974. Methane production in the interstitial waters of sulfate-depleted marine sediments. *Science*, **185**(4157), 1167-1169.

Mason, O.U., Case, D.H., Naehr, T.H., Lee, R.W., Thomas, R.B., Bailey, J.V., Orphan, V.J. 2015. Comparison of archaeal and bacterial diversity in methane seep carbonate nodules and host sediments, Eel River Basin and Hydrate Ridge, USA. *Microbiol. Ecol.*, **70**(3), 766-784.

McGlynn, S.E., Chadwick, G.L., Kempes, C.P., Orphan, V.J. 2015. Single cell activity reveals direct electron transfer in methanotrophic consortia. *Nature*, **526**(7574), 531-535.

Merkel, A.Y., Huber, J.A., Chernyh, N.A., Bonch-Osmolovskaya, E.A., Lebedinsky, A.V. 2013. Detection of putatively thermophilic anaerobic methanotrophs in diffuse hydrothermal vent fluids. *Appl. Environ. Microbiol.*, **79**(3), 915-923.

Meulepas, R.J.W., Jagersma, C.G., Gieteling, J., Buisman, C.J.N., Stams, A.J.M., Lens, P.N.L. 2009a. Enrichment of anaerobic methanotrophs in sulfate-reducing membrane bioreactors. *Biotechnol. Bioeng.*, **104**(3), 458-470.

Meulepas, R.J.W., Jagersma, C.G., Khadem, A., Stams, A.J.W., Lens, P.N.L. 2010a. Effect of methanogenic substrates on anaerobic oxidation of methane and sulfate reduction by an anaerobic methanotrophic enrichment. *Appl. Microbiol. Biotechnol.*, **87**(4), 1499-1506.

Meulepas, R.J.W., Jagersma, C.G., Khadem, A.F., Buisman, C.J.N., Stams, A.J.M., Lens, P.N.L. 2009b. Effect of environmental conditions on sulfate reduction with methane as electron donor by an Eckernförde Bay enrichment. *Environ. Sci. Technol.*, **43**(17), 6553-6559.

Meulepas, R.J.W., Jagersma, C.G., Zhang, Y., Petrillo, M., Cai, H., Buisman, C.J.N., Stams, A.J.W., Lens, P.N.L. 2010b. Trace methane oxidation and the methane dependency of sulfate reduction in anaerobic granular sludge. *FEMS Microbiol. Ecol.*, **72**(2), 261-271.

Meulepas, R.J.W., Stams, A.J.M., Lens, P.N.L. 2010c. Biotechnological aspects of sulfate reduction with methane as electron donor. *Rev. Environ. Sci. Biotechnol.*, **9**(1), 59-78.

Meyerdierks, A., Kube, M., Kostadinov, I., Teeling, H., Glöckner, F.O., Reinhardt, R., Amann, R. 2010. Metagenome and mRNA expression analyses of anaerobic methanotrophic archaea of the ANME-1 group. *Environ. Microbiol.*, **12**(2), 422-439.

Michaelis, W., Seifert, R., Nauhaus, K., Treude, T., Thiel, V., Blumenberg, M., Knittel, K., Gieseke, A., Peterknecht, K., Pape, T., Boetius, A., Amann, R., Jørgensen, B.B., Widdel, F., Peckmann, J., Pimenov, N.V., Gulin, M.B. 2002. Microbial reefs in the Black Sea fueled by anaerobic oxidation of methane. *Science*, **297**(5583), 1013-1015.

Milucka, J., Ferdelman, T.G., Polerecky, L., Franzke, D., Wegener, G., Schmid, M., Lieberwirth, I., Wagner, M., Widdel, F., Kuypers, M.M.M. 2012. Zero-valent sulphur is a key intermediate in marine methane oxidation. *Nature*, **491**(7425), 541-546.

Miyashita, A., Mochimaru, H., Kazama, H., Ohashi, A., Yamaguchi, T., Nunoura, T., Horikoshi, K., Takai, K., Imachi, H. 2009. Development of 16S rRNA gene-targeted primers for detection of archaeal anaerobic methanotrophs (ANMEs). *FEMS Microbiol. Lett.*, **297**(1), 31-37.

Moran, J.J., Beal, E.J., Vrentas, J.M., Orphan, V.J., Freeman, K.H., House, C.H. 2008. Methyl sulfides as intermediates in the anaerobic oxidation of methane. *Environ. Microbiol.*, **10**(1), 162-173.

Musat, N., Halm, H., Winterholler, B., Hoppe, P., Peduzzi, S., Hillion, F., Horreard, F., Amann, R., Jørgensen, B.B., Kuypers, M.M.M. 2008. A single-cell view on the ecophysiology of anaerobic phototrophic bacteria. *Proc. Natl. Acad. Sci. USA*, **105**(46), 17861-17866.

Musat, N., Musat, F., Weber, P.K., Pett-Ridge, J. 2016. Tracking microbial interactions with NanoSIMS. *Curr. Opin. Biotech.*, **41**(7), 114-121.

Muyzer, G., Stams, A.J. 2008. The ecology and biotechnology of sulphate-reducing bacteria. *Nat. Rev. Microbiol.*, **6**(6), 441-454.

Nauhaus, K., Albrecht, M., Elvert, M., Boetius, A., Widdel, F. 2007. *In vitro* cell growth of marine archaeal-bacterial consortia during anaerobic oxidation of methane with sulfate. *Environ. Microbiol.*, **9**(1), 187-196.

Nauhaus, K., Boetius, A., Kruger, M., Widdel, F. 2002. *In vitro* demonstration of anaerobic oxidation of methane coupled to sulphate reduction in sediment from a marine gas hydrate area. *Environ. Microbiol.*, **4**(5), 296-305.

Nauhaus, K., Treude, T., Boetius, A., Kruger, M. 2005. Environmental regulation of the anaerobic oxidation of methane: a comparison of ANME-I and ANME-II communities. *Environ. Microbiol.*, **7**(1), 98-106.

Nazaries, L., Murrell, J.C., Millard, P., Baggs, L., Singh, B.K. 2013. Methane, microbes and models: fundamental understanding of the soil methane cycle for future predictions. *Environ. Microbiol.*, **15**(9), 2395-2417.

Nielsen, J., Nielsen, P. 2010. Combined microautoradiography and fluorescence *in situ* hybridization (MAR-FISH) for the identification of metabolically active microorganisms. in: *Handbook of Hydrocarbon and Lipid Microbiology*, Vol. 1, Springer Berlin Heidelberg, Germany, pp. 4093-4102.

Niemann, H., Duarte, J., Hensen, C., Omoregie, E., Magalhaes, V.H., Elvert, M., Pinheiro, L.M., Kopf, A., Boetius, A. 2006a. Microbial methane turnover at mud volcanoes of the Gulf of Cadiz. *Geochim. Cosmochim. Ac.*, **70**(21), 5336-5355.

Niemann, H., Elvert, M. 2008. Diagnostic lipid biomarker and stable carbon isotope signatures of microbial communities mediating the anaerobic oxidation of methane with sulphate. *Org. Geochem.*, **39**(12), 1668-1677.

Niemann, H., Elvert, M., Hovland, M., Orcutt, B., Judd, A., Suck, I., Gutt, J., Joye, S., Damm, E., Finster, K., Boetius, A. 2005. Methane emission and consumption at a North Sea gas seep (Tommeliten area). *Biogeosciences*, **2**(4), 335-351.

Niemann, H., Losekann, T., de Beer, D., Elvert, M., Nadalig, T., Knittel, K., Amann, R., Sauter, E.J., Schluter, M., Klages, M., Foucher, J.P., Boetius, A. 2006b. Novel microbial communities of the Haakon Mosby mud volcano and their role as a methane sink. *Nature*, **443**(7113), 854-858.

Novikova, S.A., Shnyukov, Y.F., Sokol, E.V., Kozmenko, O.A., Semenova, D.V., Kutny, V.A. 2015. A methane-derived carbonate build-up at a cold seep on the Crimean slope, north-western Black Sea. *Mar. Geol.*, **363**(8), 160-173.

Ohtomo, Y., Ijiri, A., Ikegawa, Y., Tsutsumi, M., Imachi, H., Uramoto, G.-I., Hoshino, T., Morono, Y., Sakai, S., Saito, Y., Tanikawa, W., Hirose, T., Inagaki, F. 2013. Biological $CO_2$ conversion to acetate in subsurface coal-sand formation using a high-pressure reactor system. *Front. Microbiol.*, **4**(361), 1-17.

Orcutt, B., Boetius, A., Elvert, M., Samarkin, V., Joye, S.B. 2005. Molecular biogeochemistry of sulfate reduction, methanogenesis and the anaerobic oxidation of methane at Gulf of Mexico cold seeps. *Geochim. Cosmochim. Ac.*, **69**(17), 4267-4281.

Orcutt, B., Samarkin, V., Boetius, A., Joye, S. 2008. On the relationship between methane production and oxidation by anaerobic methanotrophic communities from cold seeps of the Gulf of Mexico. *Environ. Microbiol.*, **10**(5), 1108-1117.

Orphan, V.J., Hinrichs, K.-U., Ussler, W., Paull, C.K., Taylor, L.T., Sylva, S.P., Hayes, J.M., DeLong, E.F. 2001a. Comparative analysis of methane-oxidizing archaea and sulfate-reducing bacteria in anoxic marine sediments. *Appl. Environ. Microbiol.*, **67**(4), 1922-1934.

Orphan, V.J., House, C.H. 2009. Geobiological investigations using secondary ion mass spectrometry: microanalysis of extant and paleo-microbial processes. *Geobiology*, **7**(3), 360-372.

Orphan, V.J., House, C.H., Hinrichs, K.-U., McKeegan, K.D., DeLong, E.F. 2001b. Methane-consuming archaea revealed by directly coupled isotopic and phylogenetic analysis. *Science*, **293**(5529), 484-487.

Orphan, V.J., House, C.H., Hinrichs, K.-U., McKeegan, K.D., DeLong, E.F. 2002. Multiple archaeal groups mediate methane oxidation in anoxic cold seep sediments. *Proc. Natl. Acad. Sci. USA*, **99**(11), 7663-7668.

Orphan, V.J., Ussler, W., Naehr, T.H., House, C.H., Hinrichs, K.-U., Paull, C.K. 2004. Geological, geochemical, and microbiological heterogeneity of the seafloor around methane vents in the Eel River Basin, offshore California. *Chem. Geol.*, **205**(3-4), 265-289.

Pachiadaki, M., G. , Lykousis, V., Stefanou, E., G., Kormas, K., A. 2010. Prokaryotic community structure and diversity in the sediments of an active submarine mud volcano (Kazan mud volcano, East Mediterranean Sea). *FEMS Microbiol. Ecol.*, **72**(3), 429-444.

Pachiadaki, M.G., Kallionaki, A., Dählmann, A., De Lange, G.J., Kormas, K.A. 2011. Diversity and spatial distribution of prokaryotic communities along a sediment vertical profile of a deep-sea mud volcano. *Microb. Ecol.*, **62**(3), 655-668.

Pancost, R.D., Hopmans, E.C., Damste, J.S.S. 2001. Archaeal lipids in Mediterranean cold seeps: Molecular proxies for anaerobic methane oxidation. *Geochim. Cosmochim. Ac.*, **65**(10), 1611-1627.

Parkes, R.J., Cragg, B.A., Banning, N., Brock, F., Webster, G., Fry, J.C., Hornibrook, E., Pancost, R.D., Kelly, S., Knab, N., Jørgensen, B.B., Rinna, J., Weightman, A.J. 2007. Biogeochemistry and biodiversity of methane cycling in subsurface marine sediments (Skagerrak, Denmark). *Environmental Microbiology*, **9**(5), 1146-1161.

Parkes, R.J., Cragg, B.A., Wellsbury, P. 2000. Recent studies on bacterial populations and processes in subseafloor sediments: A review. *Hydrogeol. J.*, **8**(1), 11-28.

Paul, J., Scholin, C., Van Den Engh, G., Perry, M.J. 2007. *In situ* instrumentation. *Oceanography*, **20**(2), 70-78.

Pernthaler Annelie , P.J., Amann Rudolf. 2002. Fluorescence *in situ* hybridization and catalyzed reporter deposition for the identification of marine bacteria. *Applied and Environ. Microbiol.*, **68**(6), 3094-3101.

Pinero, E., Marquardt, M., Hensen, C., Haeckel, M., Wallmann, K. 2013. Estimation of the global inventory of methane hydrates in marine sediments using transfer functions. *Biogeosciences*, **10**(2), 959-975.

Pohlman, J.W., Ruppel, C., Hutchinson, D.R., Downer, R., Coffin, R.B. 2008. Assessing sulfate reduction and methane cycling in a high salinity pore water system in the northern Gulf of Mexico. *Mar. Petrol. Geol.*, **25**(9), 942-951.

Polerecky, L., Adam, B., Milucka, J., Musat, N., Vagner, T., Kuypers, M.M. 2012. Look@ NanoSIMS–a tool for the analysis of nanoSIMS data in environmental microbiology. *Environ. Microbiol.*, **14**(4), 1009-1023.

Poser, A., Lohmayer, R., Vogt, C., Knoeller, K., Planer-Friedrich, B., Sorokin, D., Richnow, H.H., Finster, K. 2013. Disproportionation of elemental sulfur by haloalkiphilic bacteria from soda lakes. *Extremophiles* **17**(6), 1003-1012.

Preston, C.M., Harris, A., Ryan, J.P., Roman, B., Marin III, R., Jensen, S., Everlove, C., Birch, J., Dzenitis, J.M., Pargett, D. 2011. Underwater application of quantitative PCR on an ocean mooring. *PLoS ONE*, **6**(8), e22522.

Pruesse, E., Peplies, J., Glöckner, F.O. 2012. SINA: accurate high-throughput multiple sequence alignment of ribosomal RNA genes. *Bioinformatics*, **28**(14), 1823-1829.

Rabaey, K., Rozendal, R.A. 2010. Microbial electrosynthesis-revisiting the electrical route for microbial production. *Nat. Rev. Microbiol.*, **8**(10), 706-716.

Raghoebarsing, A.A., Pol, A., van de Pas-Schoonen, K.T., Smolders, A.J.P., Ettwig, K.F., Rijpstra, W.I.C., Schouten, S., Damsté, J.S.S., Op den Camp, H.J.M., Jetten, M.S.M., Strous, M. 2006. A microbial consortium couples anaerobic methane oxidation to denitrification. *Nature*, **440**(7086), 918-921.

Reeburgh, W.S. 1980. Anaerobic methane oxidation: rate depth distributions in Skan Bay sediments. *Earth Planet. Sci. Lett.*, **47**(3), 345-352.

Reeburgh, W.S. 1976. Methane consumption in Cariaco Trench waters and sediments. *Earth Planet. Sci. Lett.*, **28**(3), 337-344.

Reeburgh, W.S. 2007. Oceanic methane biogeochemistry. *Chem. Rev.*, **107**(2), 486-513.

Reguera, G., McCarthy, K.D., Mehta, T., Nicoll, J.S., Tuominen, M.T., Lovley, D.R. 2005. Extracellular electron transfer via microbial nanowires. *Nature*, **435**(7045), 1098-1101.

Reitner, J., Peckmann, J., Reimer, A., Schumann, G., Thiel, V. 2005. Methane-derived carbonate build-ups and associated microbial communities at cold seeps on the lower Crimean shelf (Black Sea). *Facies*, **51**(1-4), 66-79.

Riedinger, N., Formolo, M.J., Lyons, T.W., Henkel, S., Beck, A., Kasten, S. 2014. An inorganic geochemical argument for coupled anaerobic oxidation of methane and iron reduction in marine sediments. *Geobiology*, **12**(2), 172-181.

Rinke, C., Lee, J., Nath, N., Goudeau, D., Thompson, B., Poulton, N., Dmitrieff, E., Malmstrom, R., Stepanauskas, R., Woyke, T. 2014. Obtaining genomes from uncultivated environmental microorganisms using FACS–based single-cell genomics. *Nat. Protoc.*, **9**(5), 1038-1048.

Roalkvam, I., Dahle, H., Chen, Y., Jørgensen, S.L., Haflidason, H., Steen, I.H. 2012. Fine-scale community structure analysis of ANME in Nyegga sediments with high and low methane flux. *Front. Microbiol.*, **3**(216), 1-13.

Roalkvam, I., Jørgensen, S.L., Chen, Y., Stokke, R., Dahle, H., Hocking, W.P., Lanzén, A., Haflidason, H., Steen, I.H. 2011. New insight into stratification of anaerobic methanotrophs in cold seep sediments. *FEMS Microbiol. Ecol.*, **78**(2), 233-243.

Rossel, P.E., Lipp, J.S., Fredricks, H.F., Arnds, J., Boetius, A., Elvert, M., Hinrichs, K.-U. 2008. Intact polar lipids of anaerobic methanotrophic archaea and associated bacteria. *Org. Geochem*, **39**(8), 992-999.

Rotaru, A.-E., Shrestha, P.M., Liu, F., Markovaite, B., Chen, S., Nevin, K.P., Lovley, D.R. 2014. Direct interspecies electron transfer between *Geobacter metallireducens* and *Methanosarcina barkeri*. *Appl. Environ. Microbiol.*, **80**(15), 4599-4605.

Rudolf K, T. 2011. Anaerobic oxidation of methane with sulfate: on the reversibility of the reactions that are catalyzed by enzymes also involved in methanogenesis from $CO_2$. *Curr. Opin. Microbiol*, **14**(3), 292-299.

Ruff, S.E., Biddle, J.F., Teske, A.P., Knittel, K., Boetius, A., Ramette, A. 2015. Global dispersion and local diversification of the methane seep microbiome. *Proc. Natl. Acad. Sci. USA*, **112**(13), 4015-4020.

Saitou, N., Nei, M. 1987. The neighbor-joining method: a new method for reconstructing phylogenetic trees. *Mol. Biol. Evol.*, **4**(4), 406-425.

Sauer, P., Glombitza, C., Kallmeyer, J. 2012. A system for incubations at high gas partial pressure. *Front. Microbiol.*, **3**(25), 225-233.

Scheller, S., Goenrich, M., Boecher, R., Thauer, R.K., Jaun, B. 2010. The key nickel enzyme of methanogenesis catalyses the anaerobic oxidation of methane. *Nature*, **465**(7298), 606-608.

Scheller, S., Yu, H., Chadwick, G.L., McGlynn, S.E., Orphan, V.J. 2016. Artificial electron acceptors decouple archaeal methane oxidation from sulfate reduction. *Science*, **351**(6274), 703-707.

Scholin, C., Doucette, G., Jensen, S., Roman, B., Pargett, D., Marin III, R., Preston, C., Jones, W., Feldman, J., Everlove, C. 2009. Remote detection of marine microbes, small invertebrates, harmful algae, and biotoxins using the environmental sample processor (ESP). *Oceanography*, **22**(2), 158-167.

Schreiber, L., Holler, T., Knittel, K., Meyerdierks, A., Amann, R. 2010. Identification of the dominant sulfate-reducing bacterial partner of anaerobic methanotrophs of the ANME-2 clade. *Environ. Microbiol.*, **12**(8), 2327-2340.

Schubert, C.J., Coolen, M.J.L., Neretin, L.N., Schippers, A., Abbas, B., Durisch-Kaiser, E., Wehrli, B., Hopmans, E.C., Damste, J.S.S., Wakeham, S., Kuypers, M.M.M. 2006. Aerobic and anaerobic methanotrophs in the Black Sea water column. *Environ. Microbiol.*, **8**(10), 1844-1856.

Segarra, K.E.A., Comerford, C., Slaughter, J., Joye, S.B. 2013. Impact of electron acceptor availability on the anaerobic oxidation of methane in coastal freshwater and brackish wetland sediments. *Geochim. Cosmochim. Ac.*, **115**(29), 15-30.

Segarra, K.E.A., Schubotz, F., Samarkin, V., Yoshinaga, M.Y., Hinrichs, K.-U., Joye, S.B. 2015. High rates of anaerobic methane oxidation in freshwater wetlands reduce potential atmospheric methane emissions. *Nat. Commun.*, **6**(7477),1-8.

Shi, Y., Hu, S., Lou, J., Lu, P., Keller, J., Yuan, Z. 2013. Nitrogen removal from wastewater by coupling anammox and methane-dependent denitrification in a membrane biofilm reactor. *Environ. Sci. Technol.*, **47**(20), 11577-11583.

Sievert, S.M., Kiene, R.P., Schultz-Vogt, H.N. 2007. The sulfur cycle. *Oceanography*, **20**(2), 117-123.

Sivan, O., Adler, M., Pearson, A., Gelman, F., Bar-Or, I., John, S.G., Eckert, W. 2011. Geochemical evidence for iron-mediated anaerobic oxidation of methane. *Limnol. Oceanogr.*, **56**(4), 1536-1544.

Sivan, O., Schrag, D.P., Murray, R.W. 2007. Rates of methanogenesis and methanotrophy in deep-sea sediments. *Geobiology*, **5**(2), 141-151.

Slomp, C.P., Mort, H.P., Jilbert, T., Reed, D.C., Gustafsson, B.G., Wolthers, M. 2013. Coupled dynamics of iron and phosphorus in sediments of an oligotrophic coastal basin and the impact of anaerobic oxidation of methane. *PLoS ONE*, **8**(6), e62386.

Sørensen, K., Finster, K., Ramsing, N. 2001. Thermodynamic and kinetic requirements in anaerobic methane oxidizing consortia exclude hydrogen, acetate, and methanol as possible electron shuttles. *Microb. Ecol.*, **42**(1), 1-10.

Stadnitskaia, A., Muyzer, G., Abbas, B., Coolen, M.J.L., Hopmans, E.C., Baas, M., van Weering, T.C.E., Ivanov, M.K., Poludetkina, E., Sinninghe Damste, J.S. 2005. Biomarker and 16S rDNA evidence for anaerobic oxidation of methane and related carbonate precipitation in deep-sea mud volcanoes of the Sorokin Trough, Black Sea. *Mar. Geol.*, **217**(1-2), 67-96.

Stams, A.J., Plugge, C.M. 2009. Electron transfer in syntrophic communities of anaerobic bacteria and archaea. *Nat. Rev. Microbiol.*, **7**(8), 568-577.

Straub, K.L., Schink, B. 2004. Ferrihydrite-Dependent Growth of Sulfurospirillum deleyianum through Electron Transfer via Sulfur Cycling. *Appl. Environ. Microbiol.*, **70**(10), 5744-5749.

Suarez-Zuluaga, D.A., Timmers, P.H.A., Plugge, C.M., Stams, A.J.M., Buisman, C.J.N., Weijma, J. 2015. Thiosulphate conversion in a methane and acetate fed membrane bioreactor. *Environ. Sci. Pollut. Res.*, **23**(3), 2467-2478.

Suarez-Zuluaga, D.A., Weijma, J., Timmers, P.H.A., Buisman, C.J.N. 2014. High rates of anaerobic oxidation of methane, ethane and propane coupled to thiosulphate reduction. *Environ. Sci. Pollut. Res.*, **22**(5), 3697-3704.

Suess, E. 2014. Marine cold seeps and their manifestations: geological control, biogeochemical criteria and environmental conditions. *Int. J. Earth Sci.*, **103**(7), 1889-1916.

Summers, Z.M., Fogarty, H.E., Leang, C., Franks, A.E., Malvankar, N.S., Lovley, D.R. 2010. Direct exchange of electrons within aggregates of an evolved syntrophic coculture of anaerobic bacteria. *Science*, **330**(6009), 1413-1415.

Tavormina, P.L., Ussler, W., Joye, S.B., Harrison, B.K., Orphan, V.J. 2010. Distributions of putative aerobic methanotrophs in diverse pelagic marine environments. *ISME J.*, **4**(5), 700-710.

Teske, A., Hinrichs, K.-U., Edgcomb, V., Gomez, A.D., Kysela, D., Sylva, S.P., Sogin, M.L., Jannasch, H.W. 2002. Microbial diversity of hydrothermal sediments in the Guaymas Basin: Evidence for anaerobic methanotrophic communities. *Appl. Environ. Microbiol.*, **68**(4), 1994-2007.

Thauer, R.K., Shima, S. 2008. Methane as fuel for anaerobic microorganisms. *Ann. N. Y. Acad. Sci.*, **1125**(1), 158-170.

Thiel, V., Peckmann, J., Richnow, H.H., Luth, U., Reitner, J., Michaelis, W. 2001. Molecular signals for anaerobic methane oxidation in Black Sea seep carbonates and a microbial mat. *Mar. Chem.*, **73**(2), 97-112.

Thomsen, T.R., Finster, K., Ramsing, N.B. 2001. Biogeochemical and molecular signatures of anaerobic methane oxidation in a marine sediment *Appl. Environ. Microbiol.*, **67**(4), 1646-1656.

Timmers, P.H., Gieteling, J., Widjaja-Greefkes, H.A., Plugge, C.M., Stams, A.J., Lens, P.N.L., Meulepas, R.J. 2015. Growth of anaerobic methane-oxidizing archaea and sulfate-reducing bacteria in a high-pressure membrane capsule bioreactor. *Appl. Environ. Microbiol.*, **81**(4), 1286-1296.

Treude, T., Knittel, K., Blumenberg, M., Seifert, R., Boetius, A. 2005a. Subsurface microbial methanotrophic mats in the Black Sea. *Appl. Environ. Microbiol.*, **71**(10), 6375-6378.

Treude, T., Krüger, M., Boetius, A., Jørgensen, B.B. 2005b. Environmental control on anaerobic oxidation of methane in the gassy sediments of Eckernförde Bay (German Baltic). *Limnol. Oceanogr.*, **50**(6), 1771-1786.

Treude, T., Orphan, V.J., Knittel, K., Gieseke, A., House, C.H., Boetius, A. 2007. Consumption of methane and $CO_2$ by methanotrophic microbial mats from gas seeps of the anoxic Black Sea. *Appl. Environ. Microbiol.*, **73**(7), 2271-2283.

Urmann, K., Gonzalez-Gil, G., Schroth, M.H., Hofer, M., Zeyer, J. 2004. New field method: gas push-pull test for the in-situ quantification of microbial activities in the vadose zone. *Environ. Sci. Technol.*, **39**(1), 304-310.

Ussler III, W., Preston, C., Tavormina, P., Pargett, D., Jensen, S., Roman, B., Marin III, R., Shah, S.R., Girguis, P.R., Birch, J.M. 2013. Autonomous application of quantitative PCR in the deep sea: *in situ* surveys of aerobic methanotrophs using the deep-sea environmental sample processor. *Environ. Sci. Technol.*, **47**(16), 9339-9346.

Valentine, D.L. 2002. Biogeochemistry and microbial ecology of methane oxidation in anoxic environments: a review. *Anton. Leeuw. Int. J. G.*, **81**(1-4), 271-282.

Valentine, D.L., Reeburgh, W.S., Hall, R. 2000. New perspectives on anaerobic methane oxidation. *Environ. Microbiol.*, **2**(5), 477-484.

Vigneron, A., Cruaud, P., Pignet, P., Caprais, J.-C., Cambon-Bonavita, M.-A., Godfroy, A., Toffin, L. 2013. Archaeal and anaerobic methane oxidizer communities in the Sonora Margin cold seeps, Guaymas Basin (Gulf of California). *ISME J.*, **7**(8), 1595-1608.

Vigneron, A., L'Haridon, S., Godfroy, A., Roussel, E.G., Cragg, B.A., Parkes, R.J., Toffin, L. 2015. Evidence of active methanogen communities in shallow sediments of the Sonora Margin cold seeps. *Appl. Environ. Microbiol.*, **81**(10), 3451-3459.

Wagner, M. 2009. Single-cell cophysiology of microbes as revealed by Raman microspectroscopy or secondary ion mass spectrometry imaging. *Ann. Rev. Microbiol.*, **63**(1), 411-429.

Wallmann, K., Pinero, E., Burwicz, E., Haeckel, M., Hensen, C., Dale, A., Ruepke, L. 2012. The global inventory of methane hydrate in marine sediments: A theoretical approach. *Energies*, **5**(7), 2449-2498.

Wan, M., Shchukarev, A., Lohmayer, R., Planer-Friedrich, B., Peiffer, S. 2014. Occurrence of surface polysulfides during the interaction between ferric (hydr)oxides and aqueous sulfide. *Environ. Sci. Technol.*, **48**(9), 5076-5084.

Wang, F.-P., Zhang, Y., Chen, Y., He, Y., Qi, J., Hinrichs, K.-U., Zhang, X.-X., Xiao, X., Boon, N. 2014. Methanotrophic archaea possessing diverging methane-oxidizing and electron-transporting pathways. *ISME J.*, **8**(5), 1069-1078.

Wankel, S.D., Adams, M.M., Johnston, D.T., Hansel, C.M., Joye, S.B., Girguis, P.R. 2012a. Anaerobic methane oxidation in metalliferous hydrothermal sediments: influence on carbon flux and decoupling from sulfate reduction. *Environ. Microbiol.*, **14**(10), 2726-2740.

Wankel, S.D., Huang, Y.-w., Gupta, M., Provencal, R., Leen, J.B., Fahrland, A., Vidoudez, C., Girguis, P.R. 2012b. Characterizing the distribution of methane sources and cycling in the deep sea via *in situ* stable isotope analysis. *Environ. Sci. Technol.*, **47**(3), 1478-1486.

Wankel, S.D., Joye, S.B., Samarkin, V.A., Shah, S.R., Friederich, G., Melas-Kyriazi, J., Girguis, P.R. 2010. New constraints on methane fluxes and rates of anaerobic methane oxidation in a Gulf of Mexico brine pool via *in situ* mass spectrometry. *Deep Sea Research Part II: Top. Stud. Oceanogr.*, **57**(21-23), 2022-2029.

Watrous, J.D., Dorrestein, P.C. 2011. Imaging mass spectrometry in microbiology. *Nat. Rev. Microbiol.*, **9**(9), 683-694.

Wegener, G., Bausch, M., Holler, T., Thang, N.M., Prieto Mollar, X., Kellermann, M.Y., Hinrichs, K.-U., Boetius, A. 2012. Assessing sub-seafloor microbial activity by combined stable isotope probing with deuterated water and $^{13}$C-bicarbonate. *Environ. Microbiol.*, **14**(6), 1517-1527.

Wegener, G., Krukenberg, V., Riedel, D., Tegetmeyer, H.E., Boetius, A. 2015. Intercellular wiring enables electron transfer between methanotrophic archaea and bacteria. *Nature*, **526**(7574), 587-603.

Wegener, G., Krukenberg, V., Ruff, S.E., Kellermann, M.Y., Knittel, K. 2016. Metabolic capabilities of microorganisms involved in and associated with the anaerobic oxidation of methane. *Front. Microbiol.*, **7**(46), 1-16.

Wegener, G., Niemann, H., Elvert, M., Hinrichs, K.-U., Boetius, A. 2008. Assimilation of methane and inorganic carbon by microbial communities mediating the anaerobic oxidation of methane. *Environ. Microbiol.*, **10**(9), 2287-2298.

Widdel, F., Bak, F. 1992. Gram negative mesophilic sulfate reducing bacteria. in: Balows, A., Truper, H., Dworkin, M., Harder, W., Schleifer, K. H. (Eds.), *The prokaryotes: a handbook on the biology of bacteria: ecophysiology, isolation, identification, applications.*, Vol. 2, Springer New York, USA, pp. 3352-3378.

Widdel, F., Hansen, T. 1992. The dissimilatory sulfate-and sulfur-reducing bacteria in: Balows, A., Truper, H., Dworkin, M., Harder, W., Schleifer, K. H. (Eds.), *The prokaryotes: a handbook on the biology of bacteria: ecophysiology, isolation, identification, applications.*, Vol. 1 (2nd edition), Springer New York, USA, pp. 582-624.

Wrede, C., Brady, S., Rockstroh, S., Dreier, a., Kokoschka, S., Heinzelmann, S.M., Heller, C., Reitner, J., Taviani, M., Daniel, R., Hoppert, M. 2012. Aerobic and anaerobic methane oxidation in terrestrial mud volcanoes in the Northern Apennines. *Sediment. Geol.*, **263-264**(4), 210-219.

Yamamoto, S., Alcauskas, J.B., Crozier, T.E. 1976. Solubility of methane in distilled water and seawater. *J. Chem. Eng. Data*, **21**(1), 78-80.

Yanagawa, K., Sunamura, M., Lever, M.A., Morono, Y., Hiruta, A., Ishizaki, O., Matsumoto, R., Urabe, T., Inagaki, F. 2011. Niche separation of methanotrophic archaea (ANME-1 and-2) in methane-seep sediments of the eastern Japan Sea offshore Joetsu. *Geomicrobiol. J.*, **28**(2), 118-129.

Zehnder, a.J., Brock, T.D. 1980. Anaerobic methane oxidation: occurrence and ecology. *Appl. Environ. Microbiol.*, **39**(1), 194-204.

Zehnder, A.J.B., Brock, T.D. 1979. Methane formation and methane oxidation by methanogenic bacteria. *J. Bacteriol.*, **137**(1), 420-432.

Zhang, Y., Henriet, J.-P., Bursens, J., Boon, N. 2010. Stimulation of *in vitro* anaerobic oxidation of methane rate in a continuous high-pressure bioreactor. *Biores. Technol.*, **101**(9), 3132-3138.

Zhang, Y., Maignien, L., Zhao, X., Wang, F., Boon, N. 2011. Enrichment of a microbial community performing anaerobic oxidation of methane in a continuous high-pressure bioreactor. *BMC Microbiol.*, **11**(137), 1-8.

Zhou, Z., Han, P., Gu, J.-D. 2014. New PCR primers based on mcrA gene for retrieving more anaerobic methanotrophic archaea from coastal reedbed sediments. *Appl. Microbiol. Biotechnol.*, **98**(10), 4663-4670.

# CHAPTER 3

# Microbial Sulfate-Reducing Activities in Anoxic Sediment from Marine Lake Grevelingen: Screening of Electron Donors and Acceptors

This chapter has been published as:

Bhattarai S*., Cassarini C*., Naangmenyele Z., Rene E. R., Gonzalez-Gil G., Esposito G., Lens P.N.L. (2017) Microbial sulfate-reducing activities in anoxic sediment from marine Lake Grevelingen: screening of electron donors and acceptors. *Limnology*, **19**(1), 31-41.

*Both authors have contributed equally to this paper.

**Abstract**

Sulfate-reducing bacteria in marine sediments mainly utilize sulfate as a terminal electron acceptor with different organic compounds as electron donor. This study investigated microbial sulfate reducing activity of coastal sediment from the marine Lake Grevelingen (MLG), the Netherlands using different electron donors and electron acceptors. All four electron donors (ethanol, lactate, acetate and methane) showed sulfate reducing activity with sulfate as electron acceptor, suggesting the presence of an active sulfate reducing bacterial population in the sediment, even at dissolved sulfide concentrations exceeding 12 mM. Ethanol showed the highest sulfate reduction rate of 55 $\mu$mol $g_{VSS}^{-1}$ day$^{-1}$ compared to lactate (32 $\mu$mol $g_{VSS}^{-1}$ day$^{-1}$), acetate (26 $\mu$mol $g_{VSS}^{-1}$ day$^{-1}$) and methane (4.7 $\mu$mol $g_{VSS}^{-1}$ day$^{-1}$). Sulfide production using thiosulfate and elemental sulfur as electron acceptors and methane as the electron donor was observed, however, mainly by disproportionation rather than by anaerobic oxidation of methane coupled to sulfate reduction. This study showed that the MLG sediment is capable to perform sulfate reduction by using diverse electron donors, including the gaseous and cheap electron donor methane.

**3.1 Introduction**

Microbial sulfate reduction (SR) to sulfide is a ubiquitous process in marine sediments, where it is mainly fueled by the microbial degradation of organic matter (Arndt et al. 2013) and the anaerobic oxidation of methane (AOM) (Knittel and Boetius 2009). This redox reaction is mediated by sulfate reducing bacteria (SRB) (Muyzer and Stams 2008). SRB are widely distributed and play an active role in the sulfur cycle. However, in marine sediments, they are mostly unculturable and their physiology is thus poorly described (D'Hondt et al., 2004; Xiong et al., 2013).

The microbial SR process has been successfully applied in the industry for the biological treatment of wastewater containing sulfate ($SO_4^{2-}$) or other sulfur oxyanions such as thiosulfate, sulfite or dithionite, wherein the end product sulfide can be precipitated as elemental sulfur ($S^0$) after an aerobic post-treatment or as metal sulfides in case of metal containing wastewaters (Liamleam and Annachhatre 2007; Weijma et al. 2006). Often the necessity of additional electron donors, such as ethanol or hydrogen, for the SR is expensive; therefore, it is appealing to study the activity and SR rates of SRB from diverse habitats and their performance using easily accessible and low-priced electron donors, such as methane ($CH_4$) (Gonzalez-Gil et al. 2011; Meulepas et al. 2010). The main challenge of using AOM coupled to SR (AOM-SR) as a process for the desulfurization of wastewater is the slow growth rate of the microorganisms involved (Deusner et al. 2009; Krüger et al. 2008; Meulepas et al. 2009a; Nauhaus et al. 2007; Zhang et al. 2010), which could possibly be increased by using more thermodynamically favorable sulfur compounds other than $SO_4^{2-}$, such as thiosulfate (Table 3.1) or $S^0$ which was reported to be an intermediate in the AOM induced SR (Milucka et al. 2012).

The coastal marine sediment from the marine Lake Grevelingen (MLG), the Netherlands, has a special microbial ecology as it harbors both cable bacteria (Hagens et al. 2015; Vasquez-Cardenas et al. 2015; Sulu-Gambari et al. 2016) and anaerobic methanotrophs (ANME)

(Bhattarai et al. 2017). A recent study on geochemical data modeling has predicted that SR and methanogenesis might be prominent microbial processes in the MLG sediment, while a large amount of $CH_4$ could be diffused out yielding minimum AOM (Egger et al. 2016). Nevertheless, AOM-SR was observed in the sediment in the presence of anaerobic methane oxidizing communities (Bhattarai et al. 2017). Based on these findings, high rate of SR with commonly used electron donors, such as acetate and ethanol, can be expected, while there could be possible involvement of other sulfur compounds for AOM, besides $SO_4^{2-}$, e.g. $S^0$ (Milucka et al. 2012). Therefore, the main objective of this study was to determine the sulfate reducing activities with different electron donors, i.e. ethanol, acetate and lactate in order to compare which one was preferred by the sulfate reducing communities inhabiting the sediment investigated. Further, potential involvement of alternative sulfur compounds ($S^0$ and $S_2O_3^{2-}$) as electron acceptors for AOM-SR activities were investigated and compared with the AOM-SR rate achieved by $SO_4^{2-}$ as an electron acceptor.

**Table 3.1** Reactions and standard Gibb's free energy changes at pH 7.0 ($\Delta G^{0'}$) for methane, thiosulfate, elemental sulfur, ethanol, lactate and acetate during anaerobic sulfate reduction

| Electron donor | Reaction | $\Delta G^{0'}$ [a] kJ mol$^{-1}$ electron donor |
|---|---|---|
| Methane | $CH_4 + SO_4^{2-} \rightarrow HCO_3^- + HS^- + H_2O$ | -17 |
| | $CH_4 + S_2O_3^{2-} \rightarrow HCO_3^- + 2HS^- + H^+$ | -39 |
| | $CH_4 + 4S^0 + 3H_2O \rightarrow HCO_3^- + 4HS^- + 5H^+$ | +24 |
| Thiosulfate | $S_2O_3^{2-} + H_2O \rightarrow SO_4^{2-} + HS^- + H^+$ | -22 |
| Elemental sulfur | $4S^0 + 4H_2O \rightarrow SO_4^{2-} + 3HS^- + 5H^+$ | +40 |
| Ethanol | $2CH_3CH_2OH + SO_4^{2-}$ $\rightarrow 2CH_3COO^- + HS^- + H^+ + 2H_2O$ | -32 |
| Lactate | $2CH_3CHOHCOO^- + SO_4^{2-}$ $\rightarrow 2CH_3COO^- + 2HCO_3^- + HS^- + H^+$ | -38 |
| | $3CH_3CHOHCOO^-$ $\rightarrow CH_3COO^- + 2C_2H_5COO^- + HCO_3^- + H^+$ | -169 |
| Acetate | $CH_3COO^- + SO_4^{2-} \rightarrow HS^- + 2HCO_3^-$ | - 47 |

**Note:**

[a] The $\Delta G^{0'}$ values were calculated from Gibbs free energies of formation from the elements at standard temperature and pressure, as obtained from Thauer et al. (1977)

## 3.2 Material and methods

### 3.2.1 Study site

MLG is a former estuary which partly interacts with seawater from the North Sea by dams (Hagens et al. 2015). It receives a high input of organic matter from the North Sea during spring and summer periods. High rates of deposition and degradation of organic matter have resulted in $CH_4$ rich anoxic sediments, which, when combined with $SO_4^{2-}$ from seawater renders the site a potential niche for SR, including AOM-SR (Egger et al. 2016). The lake inhabits unique microbiota, including *Beggiatoa* mats and a novel type of *Desulfobulbus* clade "cable bacteria", in its sediment due to its seasonal hypoxia in the shallow depth and anaerobic organic rich sediment in the deeper part of the lake (Hagens et al. 2015; Sulu-Gambari et al. 2016).

### 3.2.2 Sampling

Sediment was obtained from the MLG at a water depth of 45 m from the Scharendijke Basin (51° 44.541' N; 3° 50.969' E). The sampling site has the following characteristics: salinity - 31.7 ‰, sulfate - 25 mM at the surface of the sediment which reduced up to 5 mM at deeper sediment depths (35 cm), sedimentation rate - ~3 cm $yr^{-1}$ and average temperature - 11°C (Egger et al. 2016). The sediment was anaerobic, dark colored with prominent sulfidic odor. On the vessel R/V Luctor in November 2013, coring was done by the Royal Netherlands Institute for Sea Research (Yerseke, the Netherlands). A gravity corer (UWITEC, Mondsee, Austria) was used to collect the sediments, having a core liner internal diameter of 6 cm and a length of 60 cm. The sediment core was sliced every 5 cm and the sediment layer of 10-20 cm depth (dark colored sulfidic sediment) was used for the activity tests.

### 3.2.3 Experimental Design

The wet sediment was homogenized separately in a $N_2$-purged anaerobic chamber from PLAS LABS INC$^{TM}$ and diluted with artificial seawater medium in a ratio of 1:3, and then aliquoted in 250 ml sterile serum bottles with 40 % headspace. The artificial seawater medium composed of (per liter of demineralized water): NaCl (26 g), KCl (0.5 g) $MgCl_2 \cdot 6H_2O$ (5 g), $NH_4Cl$ (0.3 g), $CaCl_2 \cdot 2H_2O$ (1.4 g), $KH_2PO_4$ (0.1 g), trace element solution (1 ml), 1 M $NaHCO_3$ (30 ml), vitamin solution (1 ml), thiamin solution (1 ml), vitamin $B_{12}$ solution (1 ml), 0.5 g $L^{-1}$ resazurin solution (1 ml) and 0.5 M $Na_2S$ solution (1 ml) (Zhang et al. 2010). The vitamins and trace element mixture was prepared according to Widdel and Bak (1992). pH was adjusted to 7.0 with sterile 1 M $Na_2CO_3$ or 1 M $H_2SO_4$ solution, which was stored under nitrogen atmosphere. The medium was kept anoxic through $N_2$ purging until the incubation with the sediment. The prepared serum bottles were incubated in the dark with gentle shaking at room temperature (~ $20 \pm 2$°C).

Activity tests were performed with different electron donors (ethanol, lactate, acetate or methane) and different electron acceptors ($SO_4^{2-}$, $S_2O_3^{2-}$ or $S^0$) along with their respective controls in duplicate. The SR activity tests were performed in triplicates, while the AOM activity tests were performed in quadruplets. The biotic and abiotic controls were prepared in duplicates for each set of experiment. The experiments for acetate (5 mM) and lactate (5 mM) were conducted for 30 days and the experiment for ethanol (5 mM) was conducted for 40 days with intermittent addition of 5 mM ethanol around 25 days. In the case of AOM-SR experiments, the incubations with $CH_4$ (2 bar) and $SO_4^{2-}$ was carried out for 225 days. Moreover, the experiments with $CH_4$ and $S_2O_3^{2-}$ or $S^0$ were conducted for 350 days. During the experiments with $CH_4$ and $S_2O_3^{2-}$ or $S^0$, the mineral medium and headspace $CH_4$ was refreshed on day 250. Almost 10 mM of electron acceptors were used in each experiment. $SO_4^{2-}$ and $S_2O_3^{2-}$ were added in the artificial seawater media as $Na_2SO_4$ (1.43 g) and $Na_2S_2O_3$ (1.58 g) as anhydrous form, both bought from Fisher Scientific (Sheepsbouwersweg, the Netherlands). $S^0$ was purchased as precipitated sulfur as powder from Fisher Scientific (Sheepsbouwersweg, the Netherlands) and homogenized in the artificial seawater medium by continuous stirring.

Wet sediment (2 ml) was withdrawn from each bottle, once every three days, for $SO_4^{2-}$ and total sulfide (TS) for all cumulative dissolved sulfide species ($H_2S$, $HS^-$ and $S^{2-}$) analysis, while the same amount of slurry was also obtained in an interval of 15 days from the batch incubations for AOM activity test with different sulfur compounds. The analysis of total dry weight and volatile suspended solids (VSS) was performed in the beginning and at the end of each sets of the experiment. In order to ascertain the quality assurance of different measurements, chemical parameters were measured in each test batch bottle. Thereafter, the average and standard deviations were estimated among the respective batch replicates.

### 3.2.4  Chemical analysis

The VSS was estimated on the basis of the difference between the dry weight total suspended solids (TSS) and the ash weight of the sediment according to the procedure outlined in Standard Methods (APHA 1995). Dissolved TS was analyzed using the methylene blue method immediately after sampling (Siegel 1965). One volume of sample (0.5 ml) was diluted to one volume of 1 M NaOH to raise the pH to prevent the volatilization of sulfide. $SO_4^{2-}$ was analyzed using an Ion Chromatograph system (Dionex-ICS-1000 with AS-DV sampler), as described previously (Villa-Gomez et al. 2011). The pH was measured using a pH indicator paper.

### 3.2.5  Rate calculations

The volumetric SR and TS production rates were calculated as described in Eq. 3.1 to Eq. 3.4 (Meulepas et al. 2009a):

$$\text{Volumetric sulfate reduction rate} = \frac{[SO_4^{2-}{}_{(t)}]-[SO_4^{2-}{}_{(t+\Delta t)}]}{\Delta t} \qquad \text{Eq. 3.1}$$

$$\text{Volumetric sulfide production rate} = \frac{[TS_{(t)}]-[TS_{(t+\Delta t)}]}{\Delta t} \qquad \text{Eq. 3.2}$$

Where, $SO_4^{2-}{}_{(t)}$ is the concentration of $SO_4^{2-}$ at time (t) during the batch incubation, $SO_4^{2-}{}_{(t+\Delta t)}$ is the concentration of $SO_4^{2-}$ at time (t+$\Delta t$). Similarly, $TS_{(t)}$ is the concentration of TS at time (t) and $TS_{(t+\Delta t)}$ is the TS concentration at time (t+$\Delta t$). $SO_4^{2-}$/TS concentration of maximum gradient in the slope of activity test was considered for the maximum volumetric rate calculation.

$$\text{Specific sulfate reduction rate} = \frac{\text{Volumetric sulfate reduction rate}}{\text{VSS (g)}} \qquad \text{Eq. 3.3}$$

$$\text{Specific sulfide production rate} = \frac{\text{Volumetric sulfide production rate}}{\text{VSS (g)}} \qquad \text{Eq. 3.4}$$

Where, VSS is the total amount of initial VSS measured in the incubated sediment from MLG, i.e. 16.9 g.

## 3.3    Results

### 3.3.1    Sulfate reduction (SR) with ethanol, lactate and acetate as electron donor

The pH at the beginning of the experiments was ~7.5 which increased up to 8.8 towards the end of the experiments in the incubations with ethanol, lactate and acetate as electron donors. A similar trend of SR in $SO_4^{2-}$ concentration profiles was observed for the incubations with acetate and ethanol, whereas with lactate the reduction of the $SO_4^{2-}$ occurred within the first 13 days of incubation, after which the $SO_4^{2-}$ concentration remained nearly constant (Figure 3.1). Concomitant with the $SO_4^{2-}$ reduction, all incubations showed an increasing trend of dissolved TS production at the beginning and stable trend towards the end of the incubation period (Figure 3.1). Among the electron donors studied, the highest SR and TS production rates were observed in the incubation with ethanol, 55 and 78 μmol $g_{VSS}^{-1}$ day$^{-1}$, respectively (Table 3.2 and Figure 3.1b). In order to test the SR activity on the availability of electron donor and potential sulfide toxicity, 5 mM of ethanol was added to the batch incubation around day 25, after which a SR of 12 mM of $SO_4^{2-}$ was observed (Figure 3.2). Therefore, the progress of the experiment shows showed that actually ethanol was a limiting factor at that point and sulfate reduction and TS production was increased again by the addition of ethanol.

### 3.3.2    SR with CH₄ as the sole electron donor

In the batch incubations with CH₄ as the sole electron donor with different sulfur compounds, the starting pH was 7.5 and increased up to 8.5 towards the end of the activity test. Dissolved TS in the incubation with CH₄ and $SO_4^{2-}$ was around 6 mM at the end of the experiment, whilst almost 7.5 mM $SO_4^{2-}$ was consumed (Figure 3.3a). The SR rate for the incubation with CH₄ was much higher compared to the SR rate obtained in control incubations, i.e. without methane and without biomass (Figure 3.4). Similarly, the TS concentration for the incubations without the biomass and with CH₄ and $SO_4^{2-}$ was almost zero during the incubation period and for the incubation without CH₄ it was almost three times less than the cumulative TS concentration for the incubation with CH₄ and $SO_4^{2-}$. Trace organic matter utilization by the SRB might have contributed to the dissolved TS production during the initial periods (100 days) of incubation.

In the incubations with $S_2O_3^{2-}$ (Figure 3.3b) both $SO_4^{2-}$ and dissolved TS concentrations reached up to 7.8 mM and 4.2 mM, respectively during the first 50 days. After 150 days of incubations, dissolved TS increased to 6.4 mM, while $SO_4^{2-}$ was reduced from 7.8 to 1.8 mM (Figure 3.3d). After day 250, the batches were refreshed by 10 mM $S_2O_3^{2-}$ containing saline mineral medium and pressurized with 2 bar of CH₄. Then, both dissolved TS and $SO_4^{2-}$ increased exponentially to 7 mM and 6 mM, respectively, until the end of the experiment. Dissolved TS production and $SO_4^{2-}$ consumption was not observed in control incubations in abiotic incubations. The results from control incubation without CH₄ with $S_2O_3^{2-}$ showed that the SR and dissolved TS production rates were 3 times lower than those observed in the incubation with CH₄ (Figure 3.4). However, the SR rate (2.3 μmol $g_{VSS}^{-1}$ day$^{-1}$) for the control without CH₄ with $S_2O_3^{2-}$ was much higher than the SR rate (0.1 µmol $g_{VSS}^{-1}$ day$^{-1}$) for control incubation with $SO_4^{2-}$ and the absence of $CH_4$ (Figure 3.4).

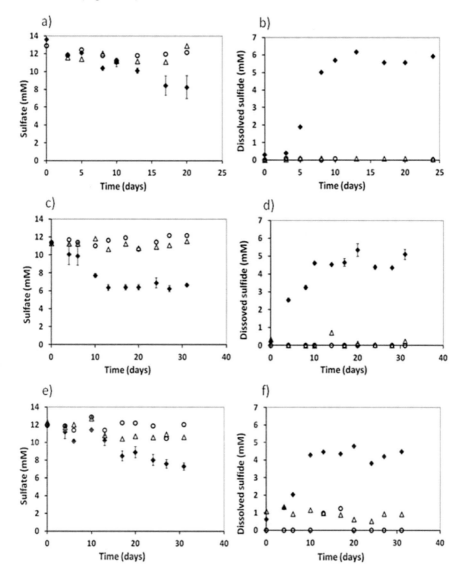

◆ Test average     O Control 1: without biomass     △ Control 2: killed biomass

**Figure 3.1** Microbial SR activity by marine Lake Grevelingen (MLG) sediment: a) sulfate consumption with ethanol, b) total sulfide production with ethanol, c) sulfate consumption with lactate, d) total sulfide production with lactate, e) sulfate consumption with acetate and f) total sulfide production with acetate.

In the incubations with $S^0$, consumption of $SO_4^{2-}$ and dissolved TS production was observed only during the first 50 days, however less in amount compared to the incubations with $SO_4^{2-}$

/$S_2O_3^{2-}$ (Figure 3.3c). Upon replacement of the mineral medium after 250 days of incubation, both $SO_4^{2-}$ and dissolved TS levels increased abruptly and reached 2.7 mM and 4.2 mM, respectively, at around day 320. After 350 days of incubation, $SO_4^{2-}$ was almost completely consumed and the dissolved TS levels increased to 6 mM. TS production and $SO_4^{2-}$ consumption was not observed in the abiotic control incubations with $S^0$. The control incubation without $CH_4$ showed a similar SR rate as in the incubation with $CH_4$ and $S^0$ (Figure 3.4).

In this study with $S_2O_3^{2-}$ and $S^0$ incubations, methane consumption was not observed; nevertheless, $CO_2$ production was almost similar for the activity test incubations with $CH_4$ and control incubation without $CH_4$. Therefore, net AOM could not be estimated when $S_2O_3^{2-}$ or $S^0$ were used as electron acceptor.

**Table 3.2** Rates of sulfate reduction (SR) and total sulfide (TS) production for  Grevelingen (MLG) sediment incubations using different electron donors and different electron acceptors

| Incubation type | SR rate | | TS  production rate | | $SO_4^{2-}$ removed | |
|---|---|---|---|---|---|---|
| | Volumetric ($\mu$mol $SO_4^{2-}$ $l^{-1}$ day$^{-1}$) | Specific ($\mu$mol $SO_4^{2-}$ $l^{-1}$ gvss$^{-1}$ day$^{-1}$) | Volumetric ($\mu$mol TS  $l^{-1}$ day$^{-1}$) | Specific ($\mu$mol TS $l^{-1}$ gvss$^{-1}$ day$^{-1}$) | mM | % |
| Ethanol + $SO_4^{2-}$ | 920 | 55 | 1320 | 78 | 6.2[a] | 90 |
| Lactate + $SO_4^{2-}$ | 540 | 32 | 580 | 34.5 | 5[a] | 88 |
| Acetate + $SO_4^{2-}$ | 440 | 26 | 560 | 33 | 5[a] | 78 |
| $CH_4$ + $SO_4^{2-}$ | 80 | 4.7 | 50 | 3 | 5.8[b] | 50 |
| $CH_4$ + $S_2O_3^{2-}$ | 120 | 7.3 | 110 | 7 | 3.3[b] | 33 |
| $CH_4$ + $S^0$ | 18 | 1 | 130 | 7.4 | 1.6[b] | 12 |

Note:

[a] The result obtained within 30 days of incubation.

[b] The result obtained within 160 days of incubation.

[c] % sulfate removed was calculated on the basis of 100 % mineralization of the added electron donor.

## 3.4 Discussion

### 3.4.1 SR by MLG sediment with different electron donors

SR by the microbiota present in MLG sediment was faster with ethanol as the substrate compared to the other electron donors tested. The SR rates with different electron donors obtained in this study were almost 100 to 200 times lower than those obtained by anaerobic granular sludge originating from bioreactors (Hao et al. 2014; Liamleam and Annachhatre 2007).

However, VSS from the sediment might overestimate the microbial biomass in the sediment, since it can include both cell biomass and organic matter present in the sediment. Therefore, SR rates determined in this study might be lower than those obtained by anaerobic granular sludge originating from bioreactors. The SR rates with ethanol, lactate and acetate (Table 3.2) were, nevertheless, higher compared to the SR rates in the *in vitro* measurements from marine coastal sediments from other shallow coastal sediments, such as Eckernförde Bay sediment with a water depth of 20 m, ranging between 0.020 and 0.465 mmol $l^{-1}$ day$^{-1}$ (Treude et al. 2005) or organic-rich shallow sediment of Limfjorden with a water depth of 10 m (eutrophic sound in Denmark connecting to the North Sea), ranging between 0.001 and 0.1 mmol $l^{-1}$ day$^{-1}$ (Jørgensen and Parkes 2010). Nevertheless, these measurements were performed in short term incubations for around five days at the *in situ* temperature ranging from 9°C to 13°C. In this study the rate was measured over a period of 20 days (for ethanol, lactate and acetate) and more than 200 days (for CH$_4$) at 20 (±2) °C.

All incubations with different electron donors (ethanol, lactate and acetate) and SO$_4^{2-}$ showed simultaneous dissolved TS production and SR. Consumption of ethanol, acetate and lactate were almost instantaneous in the different incubations and the maximum dissolved TS concentration was obtained within ~10 days of incubation (Figure 3.1). Thereafter, the maximum dissolved TS concentration remained stable which was due to the lack of electron donor as the SR activity resumed after another ethanol addition (Figure 3.2).

The microbial community in the MLG sediment was active at dissolved TS concentrations exceeding 10 mM. SRB have a wide range of TS tolerance, up to ~16 mM of dissolved TS present in the incubation medium (Reis et al. 1992). A detailed study of dissolved TS toxicity onto marine anaerobic AOM-SR consortia is still lacking. Nevertheless, with sediment from the Gulf of Mexico in the active seepage area, a dissolved TS concentration up to 12 mM was observed (Joye et al. 2004) and accumulation of 14 mM dissolved TS was observed in the incubation of sediment hosting AOM from hydrate ridge (Nauhaus et al. 2005). In contrast, dissolved TS toxicity was observed already at 2 mM with coastal estuarine sediment from Eckernförde bay (Meulepas et al. 2009b).

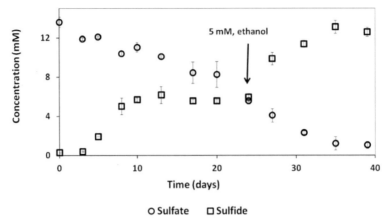

○ Sulfate    □ Sulfide

**Figure 3.2** Microbial sulfate reduction (SR) activity by marine Lake Grevelingen (MLG) sediment with ethanol showing the complete reduction of $SO_4^{2-}$ by the addition of 10 mM of total ethanol in two phases. In the starting 5 mM ethanol and 10 mM of $SO_4^{2-}$ was added to the artificial seawater medium and the incubation was spiked again with 5 mM ethanol around day 25 of incubation.

Except methane, three common sulfate reducing electron donors used in this study were known to be utilized by a wide range of SRB (Muyzer and Stams 2008). Lactate and ethanol can be fermented to short chain volatile fatty acids (VFA), such as acetate, in the presence of $SO_4^{2-}$ and then only oxidized to bicarbonate (Zellner et al. 1994). Typically, ethanol and lactate are metabolized by *Desulfovibrio, Desulfomonas, Desulfotomaculum, Desulfobulbus* (*DBB*) and *Desulfotomaculum* species of SRB to VFA or hydrogen (Muyzer and Stams 2008). Acetate is mainly utilized by *Desulfobacter, Desulfococcus, Desulfosarcina / Desulfococcus* (*DSS*), and *Desulfonema* clades of the SRB (Brock and Smith 1988). Thermodynamically, SR coupled to acetate oxidation releases the highest energy and SR in the presence of ethanol or lactate has less negative Gibb's free energy values (Table 3.1). Further, acetate is considered as major substrate for SR in marine and estuarine sediments (Parkes et al. 1989). However, the result from this study showed the lowest SR rate with acetate (Table 3.2), which suggests that acetate could have been used for other processes, such as methanogenic activity. A detailed analysis of the microbial communities along with SRB diversity could be performed in future studies to link the microbiome with their carbon source and electron donor utilization as well as their metabolic pathways.

### 3.4.2    AOM with different sulfur compounds

MLG sediment is able to utilize all three electron acceptors, i.e. $SO_4^{2-}$, $S_2O_3^{2-}$ and $S^0$, with $CH_4$ as the electron donor (Figure 3.3). While assessing the SR with $S_2O_3^{2-}/S^0$, the active sulfur disproportionated TS production was observed with these sulfur compounds instead of AOM induced TS production. $CO_2$ measurements for the incubations with $S^0$ and $S_2O_3^{2-}$ did not clearly show the oxidation of $CH_4$ to $CO_2$, as $CH_4$ remained constant and $CO_2$ was produced in both incubations with and without $CH_4$ in the headspace. Nevertheless, it was observed that the $CH_4$ consumption was ~ 5.5 mM with the simultaneous production of 1.5 mM $CO_2$ in the batch incubations with $CH_4$ and $SO_4^{2-}$ (Bhattarai et al. 2017).

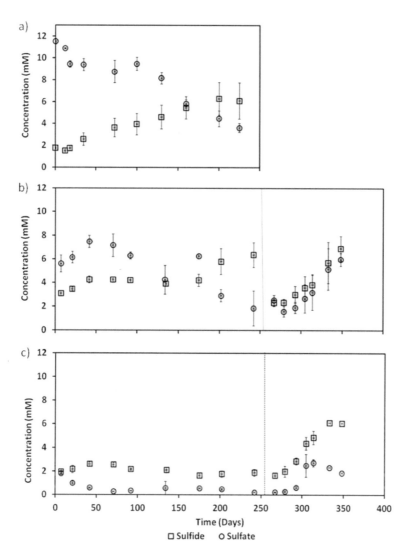

**Figure 3.3** Microbial SR activity by marine Lake Grevelingen (MLG) sediment with CH$_4$ as electron donor and a) 10 mM of SO$_4^{2-}$ (the error bar indicates standard deviation, n=4), b) 10 mM S$_2$O$_3^{2-}$ (the error bar indicates standard deviation, n=4) and c) 10 mM S$^0$ as electron acceptor (the error bar indicates standard deviation, n=3). Dashed line indicates the replacement of the medium with fresh artificial seawater medium, flush of headspace with CH$_4$ and addition of b) 10 mM S$_2$O$_3^{2-}$ and c) 10 mM S$^0$.

Similar to this study, SR rate (50 to 80 μmol SO$_4^{2-}$ l$^{-1}$ day$^{-1}$) was observed with Eckernförde Bay sediment in the beginning of an enrichment experiment in a bioreactor (Meulepas et al. 2009a). The observed dissolved TS production rate in this study with all electron acceptor were higher compared to the rate obtained after incubation at 2 bar of the cold seep sediment from Captain Aryutinov Mud Volcano (Gulf of Cadiz; 0.18 μmol TS g$_{dw}^{-1}$ day$^{-1}$) using CH$_4$ as electron donor and SO$_4^{2-}$ as an electron acceptor (Zhang et al. 2010). Similarly, a mixture of coastal sediments from the Aarhus bay and Eckernförde bay using CH$_4$ and other alkanes as electron donors with $SO_4^{2-}$ (1.5 to 2.5 μmol TS $l^{-1}$ day$^{-1}$) had lower dissolved TS production rates than those observed in this study (Suarez-Zuluaga et al. 2014).

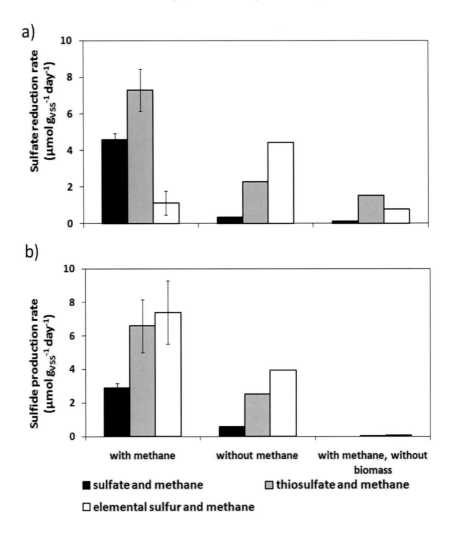

**Figure 3.4** Maximum specific a) sulfate reduction and b) total sulfide production rates (μmol $l^{-1}$ g$_{VSS}^{-1}$ day$^{-1}$) for the anaerobic methane oxidation (AOM) activity test of marine Lake Grevelingen (MLG) sediment with different electron acceptors ($SO_4^{2-}$, $S_2O_3^{2-}$ and $S^0$) and $CH_4$ as electron donor and control incubations without $CH_4$ and without biomass. The error bars indicate standard deviation (n=4 for $SO_4^{2-}$ and $S_2O_3^{2-}$) and (n=3 for $S^0$). For controls, the duplicates showed least variation so the error could not be visualized in the figure.

The SR rate obtained in the incubation with $CH_4$ and $S_2O_3^{2-}$ was comparatively higher to the SR rate obtained from the parallel incubation with $CH_4$ and $SO_4^{2-}$. Moreover, the dissolved TS production rates with $CH_4$ and $S_2O_3^{2-}$ or $S^0$ were higher (110 and 130 μmol TS $l^{-1}$ day$^{-1}$ respectively) compared to the TS production rates (80 μmol TS $l^{-1}$ day$^{-1}$ for $S_2O_3^{2-}$ and 1.2 μmol TS $l^{-1}$ day$^{-1}$ for $S^0$) of the mixture of Aarhus bay and Eckernförde bay sediment (Suarez-Zuluaga et al. 2014). The SR process with $S_2O_3^{2-}$ was indirectly activated by disproportionation rather than AOM induced SR. Previous studies indicated that the alkane degradation by marine sediments might be facilitated by the enrichment of alkane degraders with the addition of $S_2O_3^{2-}$ (Meulepas et al. 2009a; Suarez-Zuluaga et al. 2014). Further, the comparatively higher rate of SR in the control incubation without methane for the case of $S_2O_3^{2-}$ or $S^0$ might be caused by the rapid enrichment of sulfate reducing bacteria and degradation of residual organic matter. Benthic organic matter contains various fractions among which a small portion can be degraded quickly and a major portion is more recalcitrant (Arndt et al. 2013). The latter might, nevertheless, have been degraded during the long term incubation of the control without methane fueling SR in these incubations.

The highest SR rate was observed in the batch tests with $CH_4$ and $S_2O_3^{2-}$ after the first 50 days of incubations (7.3 $\mu$mol $g_{VSS}^{-1}$ day$^{-1}$), suggesting that $SO_4^{2-}$ was produced due to disproportionation and then reduced to sulfide concomitant to $CH_4$ oxidation and organic matter degradation (Figure 3.4). Theoretically more energy can be released using different oxidized forms of sulfur than $SO_4^{2-}$, such as $S_2O_3^{2-}$ (Table 3.1), which might lead to higher AOM rates (Suarez-Zuluaga et al. 2014). The fast $SO_4^{2-}$ production by $S_2O_3^{2-}$ might trigger high SRB activity and consequent high SR rate.

$S^0$ and $S_2O_3^{2-}$ are important intermediates during sulfide oxidation in marine sediments (Fossing and Jørgensen 1990). The disproportionation of $S_2O_3^{2-}$ is energetically favorable and the disproportionation of $S^0$ requires energy unless an oxidant, as Fe (III), renders the reaction more energetically favorable (Finster 2008) or in alkaline environments, such as soda lakes (Poser et al. 2013). Similarly, high rates of $S_2O_3^{2-}$ disproportionation were reported in a study with coastal marine sediment and $CH_4$ as the sole carbon source (Suarez-Zuluaga et al. 2015). In that study, a high number of *Desulfocapsa* was observed, which are specialized in disproportionation of sulfur compounds (Finster et al. 1998). Further, pH of the sediment from MLG ranged between 7.6 and 8.4 (Hagens et al. 2015), while the amount of Fe oxides in the sediment ranged between 20 and 50 $\mu$mol g$^{-1}$ (Egger et al. 2016). Therefore, microbial $S^0$ disproportionation might have been possible due to Fe (III) acting as sulfide scavenger, e.g. by bacteria such as *Desulfocapsa* (Finster 2008) as majority of these disproportionating bacteria need Fe (III) or Mn (IV) as sulfide scavenger. Alternatively, $S^0$ disproportionation could have been via the metabolism of haloalkaliphilic bacteria, which can disproportionate $S^0$ without Fe(III) or Mn(IV). However, these bacteria are commonly found in soda lakes (Poser et al. 2013) and likely do not occur in the MLG sediment. In order to decipher among these mechanisms, the archaeal and bacterial community inhabiting the MLG sediment should be further studied by e.g. genome analysis and fluorescence *in situ* Hhbridization (FISH) techniques as catalyzed reporter deposition-FISH (CARD-FISH) and FISH with micro radiography (FISH-MAR).

$CH_4$ could be utilized by the microbial community present in the MLG sediment (Bhattarai et al. 2017). The MLG sediment is capable to perform SR by using diverse electron donors, including the gaseous and cheap electron donor methane. $S^0$ and $S_2O_3^{2-}$ as electron acceptors showed sulfur disproportionation possibly by SRB, however, $S_2O_3^{2-}$ could be used as a trigger for faster SR. Further, the $SO_4^{2-}$ reducing microbial community in the studied sediment is active at high TS concentrations (Figure 3.2) and they are comparable to previously studied methane seep sediments (Zhang et al. 2010). Therefore, the MLG sediment can be used for further enrichment in bioreactors. Moreover, investigations of the SRB and ANME species responsible for AOM-SR and of sulfur disproportionation are required to exploit their potential application in the field of environmental biotechnology.

## 3.5 Conclusion

This study aimed to explore microbial sulfate reducing activity of coastal sediment from the MLG using different electron donors and electron acceptors, including gaseous electron donor i.e. $CH_4$. The $SO_4^{2-}$ reducing microbial community in the sediment was capable to utilize all four different tested electron donors namely, ethanol, lactate, acetate and $CH_4$. Moreover, when $S_2O_3^{2-}$ and $S^0$ were supplied along with $CH_4$ to incubations instead of $SO_4^{2-}$, SR activity was observed in all the cases, mainly by disproportionation of sulfur compound, rather than AOM. Among the three tested sulfur compounds used as an electron acceptor with methane, the higher SR rate was observed with $S_2O_3^{2-}$, though via disproportionation. Therefore, the use $S_2O_3^{2-}$ in the initial phase of bioreactor operation may activate the faster rate of SR. Thus, this study widens our understanding on potential use of marine sediments with diverse microbial clade in $SO_4^{2-}$ reducing waste water treating bioreactors which can be supplied with cheaper gaseous electron donor such as methane.

## 3.6 References

APHA. 1995. Standard methods for the examination of water and wastewater. *American Public Health Association*, (19th edition), Washington DC, USA. pp.1325.

Arndt, S., Jørgensen, B.B., LaRowe, D.E., Middelburg, J., Pancost, R., Regnier, P. 2013. Quantifying the degradation of organic matter in marine sediments: a review and synthesis. *Earth-Sci. Rev.,* **123**(8), 53-86.

Bhattarai, S., Cassarini, C., Gonzalez-Gil, G., Egger, M., Slomp, C.P., Zhang, Y., Esposito, G., Lens, P.N.L. 2017. Anaerobic methane-oxidizing microbial community in a coastal marine sediment: anaerobic methanotrophy dominated by ANME-3. *Microb. Ecol.,* **74**(3), 608-622.

Boetius, A., Ravenschlag, K., Schubert, C.J., Rickert, D., Widdel, F., Gieseke, A., Amann, R., Jørgensen, B.B., Witte, U., Pfannkuche, O. 2000. A marine microbial consortium apparently mediating anaerobic oxidation of methane. *Nature,* **407**(6804), 623-626.

Brock, T., Smith, D. 1988. Biology of Microorganisms. Prentice-Hall, Inc., New Jersey.

Bryant, M.P., Campbell, L.L., Reddy, C.A., Crabill, M.R. 1977. Growth of *Desulfovibrio* in lactate or ethanol media low in sulfate in association with $H_2$-utilizing methanogenic bacteria. *App.l Environ. Microb.,* **33**(5)1162-1169.

Daims, H., Brühl, A., Amann, R., Schleifer, K.-H., Wagner, M. 1999. The domain-specific probe EUB338 is insufficient for the detection of all Bacteria: development and evaluation of a more comprehensive probe set. *Syst. Appl. Microbiol.*, **22**(3), 434-444.

Daly, K., Sharp, R.J., McCarthy, A.J. 2000. Development of oligonucleotide probes and PCR primers for detecting phylogenetic subgroups of sulfate-reducing bacteria. *Microbiology*, **146**(7),1693-1705.

Deusner, C., Meyer, V., Ferdelman, T. 2009. High-pressure systems for gas-phase free continuous incubation of enriched marine microbial communities performing anaerobic oxidation of methane. *Biotechnol. Bioeng.* **105**(3), 524-533.

D'Hondt, S., Jørgensen, B.B., Miller, D.J., Batzke, A., Blake, R., Cragg, B.A., Cypionka, H., Dickens, G.R., Ferdelman, T., Hinrichs, K..-U. 2004. Distributions of microbial activities in deep subseafloor sediments. *Science*, **306**(5705), 2216-222.

Egger, M., Lenstra, W., Jong, D., Meysman, F.J., Sapart, C.J., van der Veen, C., Röckmann, T., Gonzalez, S., Slomp, C.P. 2016. Rapid sediment sccumulation results in high methane effluxes from coastal sediments. *PloS ONE*, **11**(8), e0161609.

Finster, K. 2008. Microbiological disproportionation of inorganic sulfur compounds. *J. Sulfur Chem.*, **29**(3-4), 281-292.

Finster, K., Liesack, W., Thamdrup, B. 1998. Elemental sulfur and thiosulfate disproportionation by *Desulfocapsa sulfoexigens* sp. nov., a new anaerobic bacterium isolated from marine surface sediment. *Appl. Environ. Microb.*, **64**(1), 119-125.

Fossing, H., Jørgensen, B.B. 1990. Oxidation and reduction of radiolabeled inorganic sulfur compounds in an estuarine sediment, Kysing Fjord, Denmark. *Geochim. Cosmochim. Ac.*, **54**(10), 2731-2742.

Gonzalez-Gil, G., Meulepas, R.J.W., Lens, P.N.L. 2011. Biotechnological aspects of the use of methane as electron donor for sulfate reduction. in: Murray, M.-Y. (Ed.), *Comprehensive Biotechnology*, Vol. 6 (2nd edition), Elsevier B.V. Amsterdam, the Netherlands, pp. 419-434.

Hagens, M., Slomp, C., Meysman, F., Seitaj, D., Harlay, J., Borges, A., Middelburg, J. 2015. Biogeochemical processes and buffering capacity concurrently affect acidification in a seasonally hypoxic coastal marine basin. *Biogeosciences*, **12**(5), 1561-1583.

Hao, T.-W., Xiang, P.-Y., Mackey, H.R., Chi, K., Lu, H., Chui, H.-K., van Loosdrecht, M.C.M., Chen, G.-H. 2014. A review of biological sulfate conversions in wastewater treatment. *Water Res.*, **65**(43), 1-21.

Heimann, A.C., Friis, A.K., Jakobsen, R. 2005. Effects of sulfate on anaerobic chloroethane degradation by an enriched culture under transient and steady-state hydrogen supply. *Water Res.*, **39**(15), 3579-3586.

Jørgensen, B.B., Parkes. R.J. 2010. Role of sulfate reduction and methane production by organic carbon degradation in eutrophic fjord sediments (Limfjorden, Denmark). *Limnol. Oceanogr.,* **53**(3), 1338-1352.

Joye, S.B., Boetius, A., Orcutt, B.N., Montoya, J.P., Schulz, H.N., Erickson, M.J., Lugo, S.K. 2004. The anaerobic oxidation of methane and sulfate reduction in sediments from Gulf of Mexico cold seeps. *Chem. Geol.,* **205**(3-4), 219-238.

Knittel, K., Boetius, A. 2009. Anaerobic oxidation of methane: progress with an unknown process. *Annu. Rev. Microbiol.,* **63**(1), 311-334.

Krüger, M., Blumenberg, M., Kasten, S., Wieland, A., Känel, L., Klock, J.-H., Michaelis, W., Seifert, R. 2008. A novel, multi-layered methanotrophic microbial mat system growing on the sediment of the Black Sea. *Environ. Microbiol.,* **10**(8), 1934-1947.

Liamleam, W., Annachhatre, A.P. 2007. Electron donors for biological sulfate reduction. *Biotechnol. Adv.,* **25**(5), 452-463.

Meulepas, R.J.W., Stams, A.J.M., Lens, P.N.L. 2010. Biotechnological aspects of sulfate reduction with methane as electron donor. *Rev. Environ. Sci. Biotechnol.,* **9**(1), 59-78.

Meulepas, R.J.W., Jagersma, C.G., Gieteling, J., Buisman, C.J.N., Stams, A.J.M., Lens, P.N.L. 2009a. Enrichment of anaerobic methanotrophs in sulfate-reducing membrane bioreactors. *Biotechnol. Bioeng.,* **104**(3), 458-470.

Meulepas, R.J.W., Jagersma, C.G., Khadem, A.F., Buisman, C.J.N., Stams, A.J.M., Lens, P.N.L. 2009b. Effect of environmental conditions on sulfate reduction with methane as electron donor by an Eckernförde Bay enrichment. *Environ. Sci. Technol.,* **43**(17), 6553-6559.

Milucka, J., Ferdelman, T.G., Polerecky, L., Franzke, D., Wegener, G., Schmid, M., Lieberwirth, I., Wagner, M., Widdel, F., Kuypers, M.M.M. 2012. Zero-valent sulphur is a key intermediate in marine methane oxidation. *Nature,* **491**(7425), 541-546.

Muyzer, G., Stams, A.J. 2008. The ecology and biotechnology of sulphate-reducing bacteria. *Nat. Rev. Microbiol.,* **6**(6), 441-454.

Nauhaus, K., Albrecht, M., Elvert, M., Boetius, A., Widdel, F. 2007. *In vitro* cell growth of marine archaeal-bacterial consortia during anaerobic oxidation of methane with sulfate. *Environ. Microbiol.,* **9**(1), 187-196.

Nauhaus, K., Treude, T., Boetius, A., Kruger, M. 2005. Environmental regulation of the anaerobic oxidation of methane: a comparison of ANME-I and ANME-II communities. *Environ. Microbiol.,* **7**(1), 98-106.

Niemann, H., Losekann, T., de Beer, D., Elvert, M., Nadalig, T., Knittel, K., Amann, R., Sauter, E.J., Schluter, M., Klages, M., Foucher, J.P., Boetius, A. 2006. Novel microbial communities of the Haakon Mosby mud volcano and their role as a methane sink. *Nature,* **443**(7113), 854-858.

Parkes, R.J., Gibson, G., Mueller-Harvey, I., Buckingham, W., Herbert, R. 1989. Determination of the substrates for sulphate-reducing bacteria within marine and esturaine sediments with different rates of sulphate reduction. *Microbiology,* **135**(1), 175-187.

Poser, A., Lohmayer, R., Vogt, C., Knoeller, K., Planer-Friedrich, B., Sorokin, D., Richnow, H.H., Finster, K. 2013. Disproportionation of elemental sulfur by haloalkiphilic bacteria from soda lakes. *Extremophiles* **17**(6), 1003-1012.

Reis, M.A.M., Almeida, J.S., Lemos, P.C., Carrondo, M.J.T. (1992) Effect of hydrogen sulfide on growth of sulfate reducing bacteria. *Biotechnol. Bioeng.,* **40**(5), 549-642.

Siegel, L.M. 1965. A direct microdetermination for sulfide. *Anal. Biochem.,* **11**(1), 126-132.

Stahl, D.A., Amann, R.I. 1991. Development and application of nucleic acid probes. In: Stackebrandt, E., Goodfellow, M. (Eds.), *Nucleic acid techniques in bacterial systematics.* John Wiley & Sons Ltd, Chichester, UK, pp. 205-248.

Suarez-Zuluaga, D.A., Timmers, P.H.A., Plugge, C.M., Stams, A.J.M., Buisman, C.J.N., Weijma, J. 2015. Thiosulphate conversion in a methane and acetate fed membrane bioreactor. *Environ. Sci. Pollut. Res.,* **23**(3), 2467-2478.

Suarez-Zuluaga, D.A., Weijma, J., Timmers, P.H.A., Buisman, C.J.N. 2014. High rates of anaerobic oxidation of methane, ethane and propane coupled to thiosulphate reduction. *Environ. Sci. Pollut. Res.,* **22**(5), 3697-3704.

Sulu-Gambari, F., Seitaj, D., Meysman, F.J., Schauer, R., Polerecky, L., Slomp, C.P. 2016. Cable bacteria control iron-phosphorus dynamics in sediments of a coastal hypoxic basin. *Environ. Sci. Technol.* **50**(3), 1227-1233.

Thauer, R.K., Jungermann, K., Decker, K. 1977. Energy conservation in chemotrophic anaerobic bacteria. *Bacteriol. Rev.,* **41**(1), 100-809.

Treude, T., Knittel, K., Blumenberg, M., Seifert, R., Boetius, A. 2005. Subsurface microbial methanotrophic mats in the Black Sea. *Appl. Environ. Microbiol.,* **71**(10), 6375-6378.

Vasquez-Cardenas, D., van de Vossenberg, J., Polerecky, L., Malkin, S.Y., Schauer, R., Hidalgo-Martinez, S., Confurius, V., Middelburg, J.J., Meysman, F.J.R., Boschker, H.T.S. 2015. Microbial carbon metabolism associated with electrogenic sulphur oxidation in coastal sediments. *ISME J.,* **9**:(9), 1966-1978.

Villa-Gomez, D., Ababneh, H., Papirio, S., Rousseau, D.P.L., Lens, P.N.L. 2011. Effect of sulfide concentration on the location of the metal precipitates in inversed fluidized bed reactors. *J. Hazard. Mater.* **192**(1), 200-207.

Widdel, F., Bak, F. 1992. Gram negative mesophilic sulfate reducing bacteria. in: Balows, A., Truper, H., Dworkin, M., Harder, W., Schleifer, K. H. (Eds.), *The prokaryotes: a handbook on the biology of bacteria: ecophysiology, isolation, identification, applications.*, Vol. 2, Springer New York, USA, pp. 3352-3378.

Widdel, F., Hansen, T. 1992. The dissimilatory sulfate-and sulfur-reducing bacteria in: Balows, A., Truper, H., Dworkin, M., Harder, W., Schleifer, K. H. (Eds.), *The prokaryotes: a handbook on the biology of bacteria: ecophysiology, isolation, identification, applications.*, Vol. 1 (2ⁿᵈ edition), Springer New York, USA, pp. 582-624.

Weijma, J., Veeken, A., Dijkman, H., Huisman, J., Lens, P.N.L. 2006. Heavy metal removal with biogenic sulphide: advancing to full-scale. in: Cervantes, F., Pavlostathis, S., van Haandel, A. (Eds.), *Advanced biological treatment processes for industrial wastewaters, principles and applications*, IWA publishing. London, pp. 321-333.

Xiong, Z.-Q., Wang, J.-F., Hao, Y.-Y., Wang, Y. 2013. Recent advances in the discovery and development of marine microbial natural products. *Mar. Drugs*, **11**(1), 700-717.

Zellner, G., Neudörfer, F., Diekmann, H. 1994. Degradation of lactate by an anaerobic mixed culture in a fluidized-bed reactor. *Water Res.*, **28**(6), 1337-1340.

Zhang, Y., Henriet, J.-P., Bursens, J., Boon, N. 2010. Stimulation of *in vitro* anaerobic oxidation of methane rate in a continuous high-pressure bioreactor. *Biores. Technol.*, **101**(9), 3132-3138.

# CHAPTER 4

# Pressure Sensitivity of ANME-3 Predominant Anaerobic Methane Oxidizing Community from Coastal Marine Lake Grevelingen Sediment

**Abstract**

Anaerobic oxidation of methane (AOM) coupled to sulfate reduction is mediated by, respectively, anaerobic methanotrophic archaea (ANME) and sulfate reducing bacteria (SRB). When a microbial community, obtained from the coastal marine Lake Grevelingen sediment and containing ANME-3 as the most abundant type of ANME, was incubated under a pressure gradient (0.1-40 MPa) for 77 days, ANME-3 appeared to be more pressure sensitive than the SRB. ANME-3 activity was higher at lower (0.1, 0.45 MPa) over higher (10, 20 and 40 MPa) $CH_4$ total pressures. Moreover, the sulfur metabolism was shifted upon changing the incubation pressure: SRB of the *Desulfobacterales* were more enriched at elevated pressures than the *Desulfubulbaceae*. This study provides evidence that ANME-3 can be constrained at shallow environments, despite the scarce bioavailable energy, because of its pressure sensitivity. Besides, the association between ANME-3 and SRB can be steered by changing solely the incubation pressure.

## 4.1   Introduction

Anaerobic oxidation of methane (AOM) coupled to sulfate reduction (SR) is a major sink in the oceanic methane ($CH_4$) budget. The net stoichiometry of this reaction is shown in Eq. 4.1 (Reeburgh, 2007):

$$CH_4 + SO_4^{2-} \rightarrow HCO_3^- + HS^- + H_2O \qquad \Delta G^{\circ'} = -16.6 \text{ kJ mol}^{-1} \text{ } CH_4 \qquad \text{Eq. 4.1}$$

The thermodynamics of this reaction depend on the concentration of dissolved $CH_4$. $CH_4$ is poorly soluble: 1.3 mM is its concentration in sea water at ambient pressure at 15°C (Yamamoto et al., 1976). Theoretically, elevated $CH_4$ partial pressures favor the AOM coupled to SR (AOM-SR) bioconversion since the Gibbs free energy becomes more negative at higher $CH_4$ partial pressures (Table 4.1), probably also stimulating the growth of the microorganisms mediating the process, namely anaerobic methanotrophs (ANME) and sulfate reducing bacteria (SRB).

ANME are grouped into three distinct clades, i.e. ANME-1, ANME-2 and ANME-3 based on the phylogenetic analysis of their 16S rRNA genes (Boetius et al., 2000a; Hinrichs et al., 1999; Knittel et al., 2005; Niemann et al., 2006). *In vitro* incubations of ANME-1 and ANME-2 dominated microbial communities from deep sea sediments in high-pressure reactors showed a strong positive relation of the activity of the microorganisms capable of the AOM-SR process with the $CH_4$ partial pressure, up to 12 MPa (Deusner et al., 2009; Krüger et al., 2005; Nauhaus et al., 2002; Zhang et al., 2010). In the ANME-2 dominated shallow marine sediment of Eckernförde Bay, the AOM-SR rate increased linearly with the $CH_4$ pressure from 0.00 to 0.15 MPa when incubated in batch, determining an affinity constant ($K_m$) for sulfate lower than 0.5 mM and a $K_m$ for $CH_4$ at least higher than 0.075 MPa (1.1 mM) (Meulepas et al., 2009b). The affinity constant ($K_m$) for $CH_4$ of ANME-2 from the Gulf of Cadiz sediment is about 37 mM (Zhang et al., 2010), which is equivalent to 3 MPa $CH_4$ partial pressure. A recent study showed that this ANME-2 dominated sediment had its optimum pressure at the *in situ* pressure (Bhattarai et al., 2018, submitted). In contrast, the $CH_4$ partial pressure influenced the growth of different subtypes of ANME-2 and SRB from the Eckernförde Bay marine sediment incubated for 240 days in batch (Timmers et al., 2015a). Thus, studying the effect of pressure on ANME and SRB will help understand the growth of the different ANME clades.

**Table 4.1** Gibbs free energy of AOM coupled to SR ($\Delta_rG'$) at different CH₄ partial pressures and assuming the following *in vitro* conditions: temperature 15°C, pH 7.0, $HCO_3^-$ 30 mM, $SO_4^{2-}$ 10 mM and $HS^-$ 0.01 mM. The maximum dissolved CH₄ concentration at a salinity of 32‰ and 15°C at different CH₄ partial pressure was determined by the Duan model (Duan et al. 2006).

| Pressure (MPa) | Concentration (mM) | $\Delta_rG'$ (KJ mol$^{-1}$) |
|---|---|---|
| 0.1 | 1.4 | -25.8 kJ mol$^{-1}$ CH₄ |
| 0.45 | 6.4 | -29.4 kJ mol$^{-1}$ CH₄ |
| 10 | 101.9 | -36.1 kJ mol$^{-1}$ CH₄ |
| 20 | 149.8 | -37.0 kJ mol$^{-1}$ CH₄ |
| 40 | 198 | -37.7 kJ mol$^{-1}$ CH₄ |

Finding ANME-SRB consortia that can grow fast at ambient pressure would be of great importance for the application of AOM-SR in the desulfurization of industrial wastewater. Sulfate and other sulfur oxyanions, such as thiosulfate, sulfite or dithionite, are contaminants discharged in fresh water by industrial activities such as food processing, fermentation, coal mining, tannery and paper processing. Biological desulfurization under anaerobic conditions is a well-known biological treatment, in which these sulfur oxyanions are anaerobically reduced to sulfide (Liamleam & Annachhatre, 2007; Sievert et al., 2007; Weijma et al., 2006). The produced sulfide precipitates with the metals, thus enabling their recovery (Meulepas et al., 2010a). In the process of groundwater, mining or inorganic wastewater desulfurization, electron donor for the sulfate reduction needs to be supplied externally. Electron donors such as ethanol, hydrogen, methanol, acetate, lactate and propionate (Liamleam & Annachhatre, 2007) are usually supplied, but these increase the operational and investment costs (Meulepas et al., 2010a). The use of easily accessible and low-priced electron donors such as CH₄ is therefore appealing for field-scale applications (Gonzalez-Gil et al., 2011). Moreover, from a logistic, economical and safety view point, bioreactors operating at ambient conditions are preferred over those operated at high pressures.

Coastal marine sediment from Lake Grevelingen (the Netherlands) hosts both ANME and SRB (Bhattarai et al., 2017). Among the ANME types, ANME-3 is predominant, which makes this sediment a beneficial inoculum to investigate the effect of pressure on ANME-3. ANME-3 is often found in cold seep areas and mud volcanoes with high CH₄ partial pressures and relatively low temperatures (Losekann et al., 2007; Niemann et al., 2006; Vigneron et al., 2013). Therefore, the shallow marine sediment from Lake Grevelingen was incubated at different $CH_4$ total pressures (0.1, 0.45, 10, 20, and 40 MPa) to study the influence of pressure on the AOM-SR activity, but also on the methanogenic activity and the potential formation of carbon (e.g. acetate, methanethiol, Valentine et al., 2000) and sulfur (e.g. elemental sulfur or polysulfides, Milucka et al., 2012) intermediates compounds. Moreover, phylogenetic analysis and microorganisms visualization by fluorescence in-situ hybridization (FISH) were used to study the community shifts in cell morphology and aggregates due to different $CH_4$ partial pressures in batch incubations of 77 days.

## 4.2 Material and methods

### 4.2.1 Site description and sampling procedure

The sediment was obtained from the Scharendijke Basin in the marine Lake Grevelingen (water depth of 45; position 51° 44.541' N; 3° 50.969' E), which is a former estuary in the southwestern part of the Netherlands. The sampling site characteristics, biochemical processes and the microbial community composition have been previously (Bahttarai et al., 2017; Egger et al., 2016; Hagens et al., 2015; Sulu-Gambari et al., 2016). Coring was done in November 2013 on the vessel R/V Luctor by the Royal Netherlands Institute for Sea Research (Yerseke, the Netherlands). The sampling procedure has been described in Chapter 3 (section 3.2.2), the sediment was kept at 4 °C in the dark in serum bottles with the headspace of $CH_4$ before until use.

### 4.2.2 Experimental design

The effect of the pressure on the $CH_4$ oxidation, SR and $CH_4$ production rate of the marine Lake Grevelingen sediment was assessed with 0.07 (± 0.01) g volatile suspended solids ($g_{VSS}$) in 200 ml pressure vessels incubated in triplicates at 0.1 MPa, 0.45 MPa (mimicking the *in situ* conditions), 10 MPa, 20 MPa and 40 MPa. The marine Lake Grevelingen sediment used as inoculum was incubated with artificial saline mineral medium with sulfate (10 mM). The vessels were flushed and pressurized with 100 % $CH_4$, from which about 20% was [13]C-labeled $CH_4$ ([13]$CH_4$). The incubation was performed at 15°C for 77 days. Two different control incubations were prepared in triplicates at 0.45 MPa: without biomass and without $CH_4$, but with nitrogen in the headspace.

Slurry samples were taken every week for chemical analysis. Approximately 1 mL sample was taken by attaching a connector and a vacuum tube to the exit port while gently opening the tap. Weight and pressure were measured in the vacuum tube before and after sampling. Pressure in each vessel was restored by adding fresh basal medium using a high performance liquid chromatography (HPLC) pump (SSI, USA).

### 4.2.3 Chemical analysis

The gas composition was measured on a gas chromatograph-mass spectrometer (GC-MS Agilent 7890A-5975C). The GC-MS system was composed of a Trace GC equipped with a

GC-GasPro column (30 m × 0.32 mm; J & W Scientific, Folsom, CA) and an Ion-Trap MS. Helium was the carrier gas at a flow rate of 1.7 ml min$^{-1}$. The column temperature was 30°C. The fractions of $CH_4$ and $CO_2$ in the headspace were derived from the peak areas in the gas chromatograph, while the fractions of $^{13}CH_4$, $^{12}CH_4$, $^{12}CO_2$ and $^{13}CO_2$ were derived from the mass spectrum as done by Shigematsu et al. (2004).

Total dissolved sulfide was measured by using the methylene blue method (Hach Lange method 8131) and a DR5000 spectrophotometer (Hach Lange GMBH, Düsseldorf, Germany). Samples for sulfate and thiosulfate analysis were first diluted in a solution of zinc acetate (5g/L) and centrifuged at 13,200g for 3 min to remove insoluble zinc sulfide, and filtrated through 0.45 µm membrane filters. Sulfate and thiosulfate concentrations were then determined by ion chromatography (Metrohm 732 IC detector) with a METROSEP A SUPP 5 - 250 column. The pH was checked by means of pH paper.

Polysulfides were methylated using the protocol by Kamyshny et al. (2006) and analyzed by reversed-phase HPLC. Elemental sulfur from the slurry sample was extracted using methanol following the method described by Kamyshny et al. (2009), but modified for small volumes. Dimethylpolysulfanes and extracted elemental sulfur were analyzed by an HPLC (HPLC 1200 Series, Agilent Technologies, USA) with diode array and multiple wavelength detector. A mixture of 90% MeOH and 10% water was used as eluent. A reversed phase C-18 column (Hypersil ODS, 125 × 4.0mm, 5 µm, Agilent Technologies, USA) was used for separation. Concentrations of dimethylpolysulfanes from $Me_2S_3$ to $Me_2S_7$ were calculated from calibration curves of polysulfides standards prepared following the protocol of Milucka et al. (2012). UV detector response to $Me_2S_8$ was calculated by the algorithm discussed in Kamyshny et al. (2004).

The VSS was estimated at the beginning of the experiment on the basis of the difference between the dry weight total suspended solids and the ash weight of the sediment according to the procedure outlined in Standard Method (APHA 1995).

### 4.2.4   Rate calculations

Both AOM and SR rates were expressed as µmol of sulfide or dissolved inorganic carbon (DIC) production per gram of VSS per day (µmol $g_{VSS}^{-1}$ $d^{-1}$). For the AOM rate calculation, the total production of $^{13}C$-carbonate species ($^{13}C$-DIC), i.e. $^{13}CO_2$ in both liquid and gas phases, $H^{13}CO_3^{-}$ and $^{13}CO_3^{2-}$ in liquid phase, were first calculated. Considering that only 20% of $CH_4$ was $^{13}CH_4$, the total $^{13}C$-DIC was divided by the fractional abundance of $^{13}C$ in the $CH_4$ measured and used for each batch to determine the total amount of DIC produced from $CH_4$ oxidation (Zhang et al., 2014). For methanogenesis and for the formation of carbonate species from other carbon sources than $CH_4$, $^{12}CH_4$ and $H^{12}CO_3^{-}$ were taken respectively, and divided by the $^{12}C$ fractional abundance. A line was plotted over the period where the decrease or increase of the different compounds ($^{12}CH_4$, $^{13}CH_4$, $H^{12}CO_3^{-}$, $H^{13}CO_3^{-}$, total dissolved sulfide and sulfate) was linear (at least four consecutive points) to estimate the rates (Meulepas et al., 2010b), which were divided by the biomass content in the vessels (0.07 ± 0.01 $g_{VSS}$ in each vessel).

The amount of $^{13}$C-DIC, $^{12}$C-DIC, $^{13}$CH$_4$ and $^{12}$CH$_4$ were calculated in µmol per pressurized vessel for each time as follows:

$$^{13}C - DIC = f^{13}CO_2 \times p \times \left( \frac{V_{gas}}{R \times T} + V_{liquid} k_{CO_2} \times \left( 1 + \frac{K_{a,CO_2}}{[H^+]} \right) \right)$$

Eq. 4.2

$$^{12}C - DIC = f^{12}CO_2 \times p \times \left( \frac{V_{gas}}{R \times T} + V_{liquid} k_{CO_2} \times \left( 1 + \frac{K_{a,CO_2}}{[H^+]} \right) \right)$$

Eq. 4.3

$$^{13}CH_4 = f^{13}CH_4 \times p \times \left( \frac{V_{gas}}{R \times T} + V_{liquid} k_{CH_4} \right)$$

Eq. 4.4

$$^{12}CH_4 = f^{12}CH_4 \times p \times \left( \frac{V_{gas}}{R \times T} + V_{liquid} k_{CH_4} \right)$$

Eq. 4.5

Nomenclature:

f = fraction from GC-MS

$V_{liquid}$ = liquid volume in each vessel in l

$V_{gas}$ = gas volume in each vessel in l

$k_{CO2}$ = Henry's law constant for $CO_2$ at sampling temperature (20°C): 0.39 mmol l$^{-1}$ kPa$^{-1}$

$k_{CH4}$ = Henry's law constant for $CH_4$ at sampling temperature (20°C): 0.0153 mmol l$^{-1}$ kPa$^{-1}$

$K_{a, CO2}$ = dissociation constant for dissolved $CO_2$: 4.7 10$^{-7}$

R = gas constant: 8.314 J$^{-1}$ mol$^{-1}$ K$^{-1}$

p = pressure in kPa

T = temperature in K

### 4.2.5 DNA extraction

DNA was extracted by using a FastDNA® SPIN Kit for soil (MP Biomedicals, Solon, OH, USA) by following the manufacturer's protocol. Approximately 0.5 g of the sediment was used for DNA extraction from the initial inoculum and ~0.5 ml of liquid obtained by washing the polyurethane foam packing with nuclease free water was used for extracting DNA from the enriched slurry. The extracted DNA was quantified and quality was checked as described previously (Bhattarai et al., 2017).

### 4.2.6 PCR amplification for 16S rRNA genes

The DNA was amplified by using the bar coded archaea specific primer pair arch-16s-V4 forward Arc516F and reverse Arc855R. The PCR reaction mixture (50 µl) contained 2 µl of DNA template (~70 ng) and other standard PCR reagents mentioned as mentioned in Bhattarai et al. (2017). PCR amplification was performed with an applied biosystem thermal cycler with a touch-down temperature program. PCR conditions consisted of a pre-denaturing step of 5 min at 95°C, followed by 10 touch-down cycles of 95°C for 30 sec, annealing at 68°C for 30 sec with a decrement per cycle to reach the optimized annealing temperature (63°C) and extension at 72°C. This was followed by 25 cycles of denaturing at 95°C for 30 sec and 30 sec of annealing and extension at 72°C. The final elongated step was extendedgfor 10 min.

The primer pairs used for bacteria were bac-16s-V4 forward bac520F 5'-3' AYT GGG YDT AAA GNG and reverse Bac802R 5'-3' TAC NNG GGT ATC TAA TCC (Song et al., 2013). The following temperature programme was used: initial denaturation step at 94°C for 5 min, followed by denaturation at 94°C for 40 sec, annealing at 42°C for 55 sec and elongation at 72°C for 40 sec (30 cycles). The final elongation step was extended to 10 min. 5 μl of the amplicons were visualized by standard agarose gel electrophoresis (1% agarose gel, a running voltage of 120 V for 30 minutes, stained by gel red) and documented using a GelDoc UV transilluminator.

After checking the correct band size, 150 μl of PCR amplicons were loaded in a 1% agarose gel and electrophoresis was performed for 120 min at 120 V. The gel bands were excised under UV light and the PCR amplicons were cleaned using E.Z.N.A.® gel extraction kit by following the manufacturer's protocol (Omega Biotek, USA).

### 4.2.7  Illumina Miseq data processing

The purified DNA amplicons were sequenced by an Illumina HiSeq 2000 (Illumina, San Diego, USA) and analyzed as detailed in Bhattarai et al. (2017). A total of 40,000 (± 20,000) sequences were assigned to archaea and bacteria each by examining the tags assigned to the amplicons. After eliminating the chimeras, sequences for archaea and bacteria, respectively, were analyzed and classified in MOTHUR (Schloss & Westcott, 2011). In short, the faulty sequences with mismatch tags or primers and with a size less than 200 bp were removed by using the shhh.flows command. Then, the putative chimeric sequences were identified and removed by the chimera.uchime command using the most abundant reads in the respective sequence data sets as references. The sequence reads were classified according to the Silva taxonomy (Pruesse et al., 2007) using the classify.seqs command and the relative abundance of each phylotype was estimated.

### 4.2.8  Quantitative real-time PCR (Q-PCR)

Archaeal and bacterial clones were used to prepare Q-PCR standard. Plasmids were isolated using the plasmid kit (Omega Biotek, USA). The plasmid was digested with the EcoR I enzyme. After digestion purification was done by gel extraction (Gel extraction Kit, Omega Biotek, USA). The copy number was calculated from the total mass and the nucleic acid concentration. Extracted DNA from the sediment at the start and at the end of the incubation period (11 weeks) was used for qPCR analysis to quantify archaea and bacteria. Amplifications were done in triplicates in a 7500 Real-Time PCR System (Applied Biosystem). Each reaction (20μl) contained 1× Power SYBR-Green PCR MasterMix (Applied Biosystems), 0.4 μM of each primer, and 5 ng template DNA. The 16S rRNA genes of bacterial origin were amplified using the primers Bac331f (5'-TCCTACGGGAGGCAGCAGT3') and Bac797r (5'-GGACTACCAGGGTCTAATCCTGTT-3') (Nadkarni et al., 2002). Cycling conditions were

95°C for 10 min; and 40 cycles at 95°C for 30 sec and 60°C for 30 sec and 72°C for 30 sec. Archaea were quantified using the primer set Arch349f (5′-GYGCASCAGKCGMGAAW-3′) and Arch806r (5′-GGACTACVSGGGTATCTAAT-3′) (Takai & Horikoshi, 2000). Cycling conditions were 95°C for 10 min; and 40 cycles at 95°C for 30 sec and 50°C for 30 sec and 72°C for 30 sec. Triplicate standard curves were obtained with 10-fold serial dilutions ranged between $10^7$ and $10^{-2}$ copies per µl of plasmids. The efficiency of the reactions was up to 100% and the $R^2$ of the standard curves were up to 0.999.

### 4.2.9   Cell visualization and counting by FISH

At the start and at the end of the incubation period (11 weeks), 200 µL of sample from each vessel was fixed in a final 2% paraformaldehyde solution for 4 h on ice. The samples were washed twice with 1× phosphate buffer saline solution (PBS). Then it was stored in a mixture of PBS and ethanol (EtOH), with a PBS/EtOH ratio of 1:1 at -20°C as previously described by (Boetius et al., 2000a). This sample was used for cell counting and FISH analysis.

100 µL of stored sample was diluted with nuclease free water and sonicated for 40 sec then filtered on 0.2 µm membrane filters. For cell counting, 200-300 µL of 20× SYBR green solution (Takara, Japan) was added on top of the filter and incubated in the dark at room temperature for 30 min. The filters were dried and mounted in a glass slide with 100 µL glycerol 10%. For FISH analysis, the filtrated sample was hybridized with archaeal probe, ARCH915 (Stahl, 1991) and bacterial probe, EUB I-III (Daims et al., 1999), with different CY3-labeled ANME probes; ANME-1 350 (Boetius et al., 2000b), ANME-2 538 (Treude et al., 2005), ANME-3 1249 (Niemann et al., 2006) and Cy5-labelled SRB specific probes; *Desulfosarcina / Desulfococcus* (DSS) DSS658 (Boetius et al., 2000a) and *Desulfobulbus* (DBB) DBB660 (Daly et al., 2000). Cells were counterstained with 4′, 6-diamidino-2-phenylindole (DAPI) (Wagner et al., 1993). The hybridization of the samples and microscopic visualization of the hybridized cells were performed as described previously (Snaidr et al., 1997).

### 4.3   Results

### 4.3.1   Conversion rates of sulfur compounds

The highest sulfide production rates of the coastal marine Lake Grevelingen sediment was in the incubations at the *in situ* pressure (0.45 MPa) and 10 MPa: 270 and 258 µmol $g_{VSS}^{-1}$ $d^{-1}$, respectively (Figure 4.1a). The sulfide production rate at 40 MPa was 109 µmol $g_{VSS}^{-1}$ $d^{-1}$, comparable to the rate with no $CH_4$ in the headspace, 99 µmol $g_{VSS}^{-1}$ $d^{-1}$ (Figure 4.1a). Similarly, high SR rates were recorded for the incubations at 0.45 MPa and 10 MPa (Figure 4.2a): 297 and 278 µmol $g_{VSS}^{-1}$ $d^{-1}$, respectively. In contrast, the SR rate at 0.1 MPa was 257 µmol $g_{VSS}^{-1}$ $d^{-1}$, while the sulfide production was only 157 µmol $g_{VSS}^{-1}$ $d^{-1}$ (Figures 4.1a and 4.2a).

Sulfide was produced in almost all the incubations, with the exception of the incubation without biomass (Figure 4.1b). The sulfate concentration profiles varied with pressure: after 40 days of incubation at 40 MPa, sulfate was not reduced anymore (Figure 4.2b). Differently, at 0.45 MPa sulfate was reduced to sulfide in a 1:1 ratio (Figure 4.3b). At 0.45 MPa, 0.98 mmol of sulfate was consumed and exactly 0.98 mmol of total dissolved sulfide was produced, closing the sulfur balance. In the incubation at 0.1 MPa, 0.37 mmol of elemental sulfur was produced along with 0.54 mmol of sulfide (Figure 4.3a). In the other incubations at different pressures, hardly any elemental sulfur was formed (Figures 4.3c-4.3f). Instead, long chain polysulfides were formed along the incubation depending on the pressure, but in small amounts ($\leq 2$ µmol per vessel2 µmol $S_6^{2-}$ per vessel was determined at 0.45 MPa $CH_4$ pressure (Figure 3b) and 1.2 and 1.4 µmol $S_6^{2-}$ per vessel at 10 MPa and 20 MPa, respectively (Appendix 1, Figures S4.1c and S4.1d).

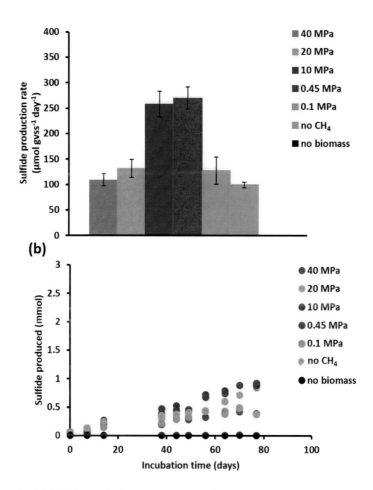

Figure 4.1 (a) Sulfide production rate and (b) sulfide concentration profiles for incubations at different pressure and controls without $CH_4$, but with $N_2$ in the headspace and without biomass. Error bars indicate the standard deviation (n=3).

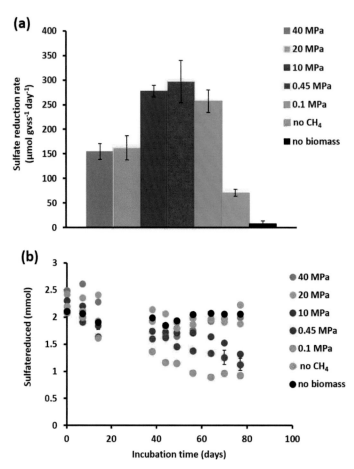

**Figure 4.2 (a)** SR rate and **(b)** sulfate concentration profiles for incubations at different pressure and controls without CH₄, but with N₂ in the headspace and without biomass. Error bars indicate the standard deviation (n=3).

### 4.3.2 AOM rates

The AOM rates were calculated from the DIC produced from $^{13}CH_4$. The $K_m$ for $CH_4$ of the marine Lake Grevelingen sediment was determined to be around 1.7 mM. The DIC production rates followed a similar trend as the sulfide production rates: the highest rate was found at 0.45 MPa and the lowest rate at 40 MPa: 320 and 38 µmol $g_{VSS}^{-1}$ $d^{-1}$, respectively (Figure 4.4a). In the incubation at 0.45 MPa, the total DIC produced from $CH_4$ was similar to the sulfide produced (Figures 4.1b and 4.4b): ~0.9 µmol per vessel. Similar trends were found for all the other incubations at different pressure, except for the vessel without $CH_4$, where only sulfide production (0.3 mmol/vessel) was recorded. However, sulfide was produced from the start for all the incubation, while the total DIC from $CH_4$ was mainly produced only after 40 days of incubation (Figures 4.1b and 4.4b). Similar trends were found for all the other incubations at different pressure, except for the vessel without $CH_4$, where only sulfide production (0.3 mmol/vessel) was recorded.

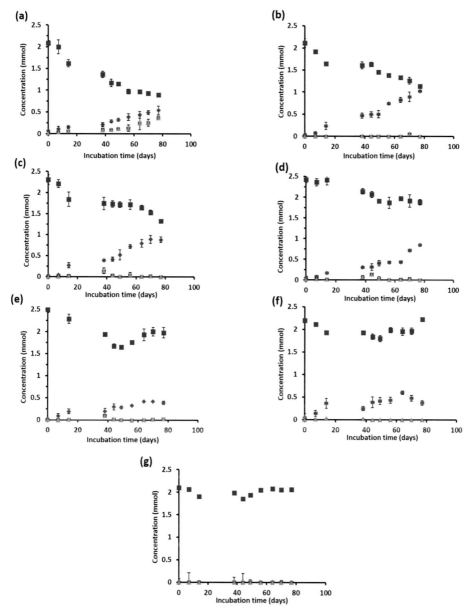

**Figure 4.3 (a)** Concentration profiles of total dissolved sulfide (◆), sulfate (■) and elemental sulfur (▲) for the incubation at **(a)** 0.1MPa, **(b)** 0.45 MPa, **(c)** 10 MPa, (d) 20 MPa, **(e)** 40 MPa, (f) without $CH_4$, and (g) without biomass. Error bars indicate the standard deviation (n=3).

### 4.3.3 Methanogenesis

CH$_4$ was produced in all the incubations, with the exception of the batches without biomass (Appendix 1, Figure S4.2). The highest amount of CH$_4$ formed was recorded in the vessel at 0.1 MPa (Appendix 1, Figure S4.2b). The highest methanogenic rate was determined in the control vessel without CH$_4$ (N$_2$ in the headspace) and at 0.1 MPa: 44 and 31 µmol g$_{VSS}$$^{-1}$ d$^{-1}$, respectively, while it was below 5 µmol g$_{VSS}$$^{-1}$ d$^{-1}$ in all the other batch incubations (Appendix 1, Figure S4.2a). Assuming that all the total $^{12}$C-DIC was produced from the oxidation of other carbon sources than CH$_4$, its production rate was low in almost all the incubations: lower than 3 µmol g$_{VSS}$$^{-1}$ d$^{-1}$, except for the incubation without CH$_4$ (64 µmol g$_{VSS}$$^{-1}$ d$^{-1}$) (Appendix 1, Figure S4.3).

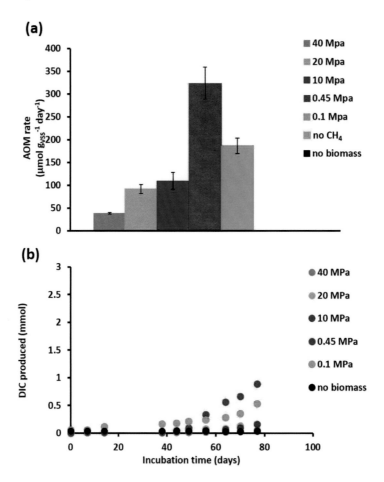

**Figure 4.4 (a)** AOM rate calculated from the linear regression over at least four successive measurements in which the calculated DIC increase over time was linear. **(b)** The DIC produced from CH$_4$ oxidation was calculated from the $^{13}$C-DIC. The AOM rate and DIC produced during AOM were determined for incubations at different pressures and controls without CH$_4$, but with N$_2$ in the headspace and without biomass. Error bars indicate the standard deviation (n=4).

#### 4.3.4  Community shifts: total cell numbers

The total bacterial and archaeal cellular numbers were accessed from Q-PCR data performed on samples after 77 days of incubations (Table 4.2). The highest increase in active cells, from 6 to $8 \times 10^7$ cells ml$^{-1}$, was found in the incubation at the *in situ* pressure of 0.45 MPa. In the incubation at 40 MPa, the amount of active total bacteria and archaea cells decreased from 6.5 to $5.8 \times 10^7$ cells ml$^{-1}$ (Table 4.2). Based on Q-PCR results, archaea grew in all the incubations, while copy numbers of bacteria decreased in the incubation without CH$_4$ and at 20 MPa. The total number of archaea increased the most in the incubation at 20 MPa (Table 4.2).

**Table 4.2** Total number of active cells and number of copies of archaea and bacteria from Q-PCR analysis per ml of wet sediment in each pressurized vessel at the start (t=0 days) and at the end of the incubation (t=77 days).

| CH$_4$ partial pressure (MPa) | Incubation time (days) | Concentration of active cells ( $\times 10^7$ cells mL$^{-1}$) | Bacteria ($\times 10^7$ copy number mL$^{-1}$) | Archaea ($\times 10^7$ copy number mL$^{-1}$) |
|---|---|---|---|---|
| 40 | 0 | $6.37 \pm 0.56$ | $3.67 \pm 0.53$ | $1.65 \pm 0.56$ |
| | 77 | $5.87 \pm 0.13$ | $3.95 \pm 0.90$ | $0.56 \pm 0.13$ |
| 20 | 0 | $6.35 \pm 0.12$ | $4.01 \pm 0.76$ | $1.07 \pm 0.12$ |
| | 77 | $6.82 \pm 0.43$ | $2.13 \pm 0.96$ | $3.11 \pm 0.43$ |
| 10 | 0 | $6.75 \pm 0.47$ | $2.21 \pm 0.06$ | $1.54 \pm 0.47$ |
| | 77 | $7.96 \pm 0.46$ | $2.45 \pm 0.78$ | $1.97 \pm 0.45$ |
| 0.45 | 0 | $5.94 \pm 0.17$ | $3.78 \pm 0.44$ | $1.62 \pm 0.17$ |
| | 77 | $8.83 \pm 0.16$ | $4.58 \pm 0.87$ | $2.21 \pm 0.16$ |
| 0.1 | 0 | $6.48 \pm 0.37$ | $4.06 \pm 0.51$ | $1.82 \pm 0.37$ |
| | 77 | $7.18 \pm 0.72$ | $4.12 \pm 0.94$ | $1.96 \pm 0.72$ |
| Without CH$_4$ | 0 | $6.22 \pm 0.39$ | $3.59 \pm 0.30$ | $1.20 \pm 0.39$ |
| | 77 | $6.02 \pm 0.35$ | $2.92 \pm 0.86$ | $1.78 \pm 0.35$ |

Based on the 16s rRNA gene analysis, both archaeal and bacterial communities were shifted along the 77 days incubation. The most abundant operational taxonomic unit (OTU) with archaeal signature are shown in Figure 4.5. Specifically, the abundance of ANME-3 among all the archaea increased the most at 0.45 and 0.1 MPa incubations, i.e. respectively three and two times more than at the start of the incubation (Figure 4.5). ANME-2a/b reads increased the most at 20 MPa, 27 times more than at the start of the incubation (Figure 4.5). Sequences of methanogens specifically belonging to the *Methanomicrobiales* were more abundant after the incubation at 0.1 MPa rather than at higher partial pressures, where *Thaumarchaeota* and *Woesearchaeaota* were more abundant in incubations at 10, 20 and even 40 MPa (Figure 4.5).

The bacterial communities were very diverse in all the incubations, the ones with the highest percentage are shown in Figure 4.6. The absolute abundance of the *Desulfobulbaceae* (DBB) as calculated from their 16s rRNA gene according to Q-PCR and Miseq results increased or remained similar at the lower pressure incubations (0.1 and 0.45 MPa), but the percentage of DBB in the total bacterial community decreased at more elevated pressures (10, 20 and 40 MPa). Differently, the absolute abundance of *Desulfobacteraceae*, as DSS, increased in all the incubations at different pressures, with the highest percentage of reads retrieved in the incubation at 20 MPa (Figure 4.6). The percentage of OTUs as assigned to *Desulfovibrio, Desulfuromonas, Halomonas* and *Sulfurovum* genes decreased in all the batch incubations (Figure 4.6).

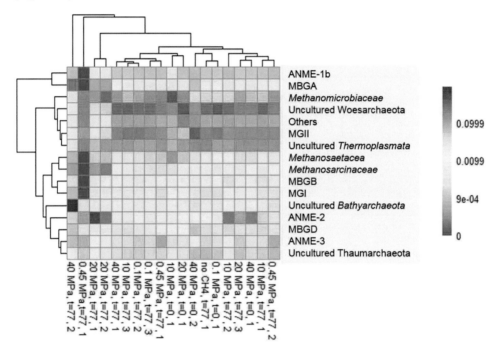

**Figure 4.5** Heat map of top most abundant 16s rRNA sequences at the beginning (t=0) and at the end of the incubations (t=77 days) of the marine Lake Grevelingen sediment at different CH$_4$ pressures and control without CH$_4$ in the headspace showing the phylogenetic affiliation up to family level as derived by high throughput sequencing of archaea.

### 4.3.5  Community shifts: FISH analysis

ANME-3 and DBB were visualized in all the batch incubations (Figures 4.7 and 4.8). At the beginning of the incubation (Figures 4.7a and 4.7b), the ANME-3 cells were preferentially visualized in aggregates with other cells. FISH images after 77 days of incubation showed variations in the aggregate morphology depending on the incubation pressure. At 0.1 and 0.45 MPa, ANME-3 was more abundant than at the beginning, while the DBB cells were not found concomitant to the ANME-3 cells (Figures 4.7d, 4.7g and 4.7h) and, even if present, the

ANME-3 outnumbered the DBB cells (Figures 4.7e, 4.7f and 4.7i). In the 10 MPa, ANME-3 was visualized more scattered and not in clusters as at the lower pressures, whereas the DBB cells were even more rarely pictured (Figures 4.8a, 4.8b, 4.8c). At 20 MPa, the ANME-3 and DBB cells were rare, however the stained cells formed tight ANME-3/DBB aggregates (Figures 4.8d, 4.8e, 4.8f). At 40 MPa, ANME-3 and DBB were the least abundant and scattered, and no aggregates could be found (Figures 4.8g, 4.8h and 4.8i).

Differently than ANME-3, more ANME-2 cells were visualized in the (77 days incubations) at higher (10, 20 and 40 MPa) than at lower (0.1 and 0.45 MPa) incubation pressures (Figures 4.9 and 4.10). DSS, the most common SRB bacterial partner of ANME-2, were most abundant at 0.1 MPa, at lower pressure they were mainly visualized together with ANME-2 (Figure 4.9). At 20 MPa only clusters of ANME-2 cells were visualized (Figures 4.10d, 4.10e and 4.10f) without DSS.

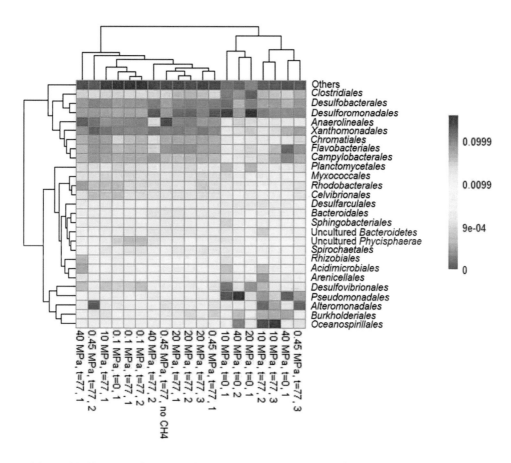

**Figure 4.6** Heat map of top most abundant 16s rRNA sequences at the beginning (t=0) and at the end of the incubations (t=77 days) of the marine Lake Grevelingen sediment at different CH₄ pressures and control without CH₄ in the headspace showing the phylogenetic affiliation up to family level as derived by high throughput sequencing of bacteria.

### 4.4 Discussion

### 4.4.1 Pressure effect on AOM in Lake Grevelingen sediment

This study showed that AOM and SR processes in Lake Grevelingen sediment depend on the the $CH_4$ partial pressure. According to Eq. 4.1, the reaction rate is expected to be stimulated by the elevated $CH_4$ partial pressure when the other parameters remain the same (Table 4.1). This expectation has been commonly accepted and has been shown in ANME-1 (Girguis et al., 2005) or ANME-2 (Meulepas et al., 2010b; Timmers et al., 2015a; Bhattarai et al., 2018 submitted) dominant communities. Figures 4.2 and 4.4 clearly illustrate the AOM-SR process of the ANME-3 dominated marine Lake Grevelingen sediment has, in contrast, an optimal pressure at 0.45 MPa among all tested conditions. This contrasts the theoretical thermodynamic calculation (Table 4.1), but is in accordance with their natural habitat, i.e. the *in situ* pressure of marine Lake Grevelingen is 0.45 MPa.

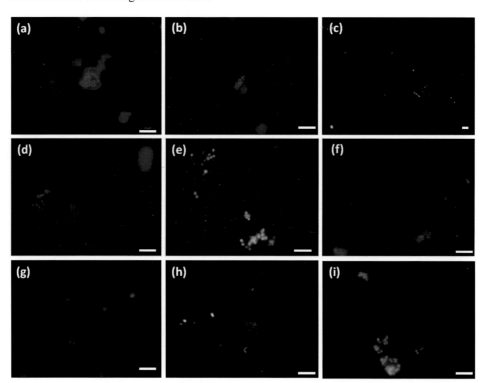

**Figure 4.7** FISH images from CY3-labeled ANME-3 in red color, CY5-labeled *Desulfobulbus* (DBB) in green and all microbial cells stained with DAPI in blue color. FISH images **(a-c)** at the beginning, and after 77 days of incubation at **(d-f)** 0.1 MPa and **(g-i)** 0.45 MPa. White scale bar representing 10 μm.

The calculated $K_m$ value on $CH_4$ based on our inoculum of around 1.7 mM is much lower than previously reported: 37 mM as calculated from an ANME-2 predominant enrichment originated from the Gulf of Cadiz (Zhang et al., 2010). Thus, the ANME cells from Grevelingen marine sediment have higher affinity for $CH_4$ than the ANME-2 from Gulf of Cadiz.

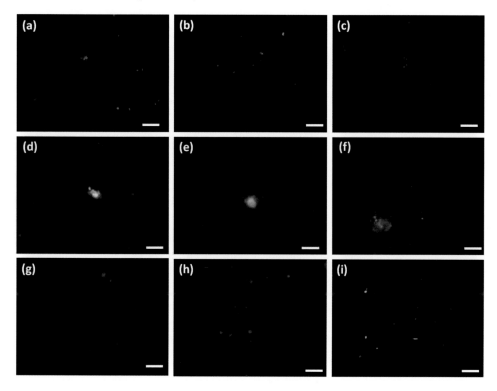

**Figure 4.8** FISH images from CY3-labeled ANME-3 in red color, CY5-labeled *Desulfobulbus* (DBB) in green and all microbial cells stained with DAPI in blue color. FISH images after 77 days of incubation at (**a-c**) 10 MPa, (**d-f**) 20 MPa and (**g-i**) 40 MPa. White scale bar representing 10 µm.

### 4.4.2 Pressure effect on ANME types

The ANME type that proliferated at lower pressure (0.1 and 0.45 MPa) was ANME-3, suggesting that the ANME-3 cells of marine Lake Grevelingen are non-piezophilic, which are easily damaged by high pressures and require extra energy to cope with the damage (Zhang et al., 2015). ANME-3 are found in cold seep areas and mud volcanoes with high $CH_4$ partial pressures and relatively low temperatures (Losekann et al., 2007; Niemann et al., 2006; Vigneron et al., 2013). However, Lake Grevelingen is a shallow sediment with high abundance of ANME-3 (Bhattarai et al., 2017) and perhaps contains different subtypes than the ones found in deep sea sediments that cannot cope with high $CH_4$ partial pressure.

The ANME-3 type is usually visualized in association with DBB as sulfate reducing partner (Losekann et al., 2007; Niemann et al., 2006). Figures 4.7 and 4.8 show that the DBB cells were not as high in number as the ANME-3 cells in any of the incubations, but they increased the most at the 0.1 MPa incubation (Figures 4.7d-4.7f). ANME-3 and DBB cells were visualized by FISH, and also through this technique the DBB cells were in general less abundant than the ANME-3. In a recent study describing the microbial ecology of Lake Grevelingen sediment (incubation pressure = 0.1 MPa), the two species could not be visualized together and the DBB cells were much less abundant than ANME-3 (Bhattarai et al., 2017), similarly to this study. At 0.1 and 0.45 MPa, ANME-3 cells were visualized in aggregates mainly detached from DBB cells (Figure 4.7). ANME-3 cells have been visualized without bacterial partner before (Omoregie et al., 2008; Vigneron et al., 2013), suggesting that this ANME type is supporting a metabolism independent of an obligatory bacterial association. In contrast, as ANME-3 and DBB decreased in number at higher pressures, most of the ANME-3 and DBB visualized at 20 MPa were forming small ANME-3/DBB clusters, suggesting that they possibly have mutual benefit at this pressure (Figure 4.8).

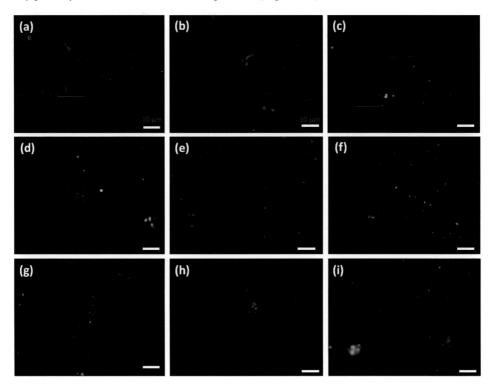

**Figure 4.9.** FISH images from CY3-labeled ANME-2 in red color, CY5-labeled *Desulforsarcina/Desulfococcus* group (DSS) in green and all microbial cells stained with DAPI in blue color. FISH images **(a-c)** at the beginning, after 77 days of incubation at **(d-f)** 0.1 MPa or **(g-i)** 0.45 MPa. White scale bar representing 10 μm.

Also sequences of ANME-2 were found by Miseq analysis (Figure 4.5) and visualized by FISH (Figures 4.9 and 4.10) in all incubations. ANME-2a/b cells were higher in number in the incubation at higher pressures (10 and 20 MPa). Also many DSS were found in all the batch incubations and as for ANME-2, they were more abundant at higher pressures (10 and 20 MPa). ANME-2 and DSS were mainly visualized in aggregates, especially at lower pressures (0.1 and

0.45 MPa). The cooperative interaction between the ANME-2 and DSS is still under debate: Milucka et al. (2012) stated that a synthrophic partner might not be required for ANME-2 and that they can be decoupled by using external electron acceptors (Scheller et al., 2016), whereas recent studies have shown direct electron transfer between the two partners (McGlynn et al., 2015; Wegener et al., 2016). Besides, the DSS might have proliferated by growth on organic carbon compound released by damaged or killed microorganisms.

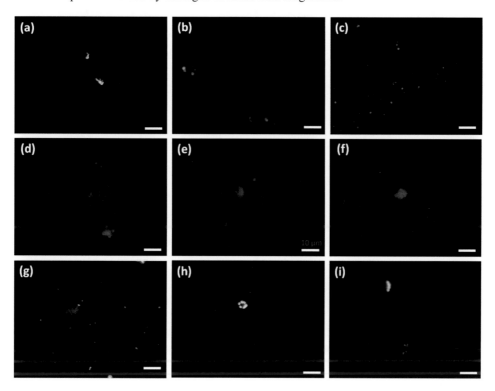

**Figure 4.10** FISH images from CY3-labeled ANME-2 in red color, CY5-labeled *Desulforsarcina/Desulfococcus* group (DSS) in green and all microbial cells stained with DAPI in blue color. FISH images after 77 days of incubation at (**a-c**) 10 MPa, (**d-f**) 20 MPa and (**g-i**) 40 MPa. White scale bar representing 10 μm.

### 4.4.3  Effect of pressure on sulfur cycle in marine Lake Grvelingen sediment

Figure 4.3 shows that the sulfur cycling in the marine Lake Grevelingen sediment community is steered by the $CH_4$ partial pressure. At 0.1 MPa $CH_4$ pressure, the reduced sulfate was converted to both sulfide and zero-valent sulfur (Figure 4.3a). The production of elemental sulfur was repressed at elevated $CH_4$ pressure (Figure 4.3), at 0.45 MPa (the incubation with the highest AOM-SR activity), the sulfur balance was closed by solely the sulfide production (Figure 4.3b). Elemental sulfur has been considered as intermediate in the SR-AOM process, which is consumed by ANME to generate energy (Milucka et al. 2012). Milucka et al. (2012) showed that ANME-2 cells could stand along without the metabolic support of the bacterial partner, assuming that $CH_4$ was oxidized to bicarbonate and sulfate was reduced to disulfide ($S_2^{2-}$) through zero-valent sulfur as an intracellular intermediate. The amount of disulfide or other polysulfides formed during the incubations (Appendix 1, Figure S4.1) was very low, in most cases below the detection limit (0.1 µmol). Further research with isotopic labeled sulfate ($^{35}S$) and nanometre scale secondary ion mass spectrometry (NanoSIMS) analysis is required to elucidate the formation of these intermediate sulfur compounds.

A shift from sulfate reducers (e.g. *Desulfobacterales*) to sulfur reducers (e.g. *Desulforomonadales*) were observed in the bacterial community from low to high $CH_4$ partial pressure (Figure 4.6). Sulfur reducing bacteria, e.g. *Desulfovibrio* or *Desulforomonas*, are more abundant at high $CH_4$ partial pressure (10, 20, 40 MPa), sulfate reducing DBB are more abundant in the incubations at lower $CH_4$ total pressure (Figure 4.7) are more abundant in the incubations at lower $CH_4$ partial pressure, where they were present in ANME-DBB aggregates and had the highest AOM-SR rates (Figures 4.2 and 4.4).

### 4.4.4 *In vitro* demonstration of SR-AOM supported ecosystem in Lake Grevelingen

This study showed that $CH_4$ and sulfate were an effective energy source supporting SR-AOM in the microbial ecosystem from the marine Lake Grevelingen sediment. Apparent *in vitro* biomass growth was observed, especially at 0.45 MPa which mimics the *in situ* pressure, with $CH_4$ and sulfate supplied as the sole energy sources (Table 4.2). At incubation conditions similar to *in situ* conditions (p = 0.45 MPa, T = 15°C, pH = 7), the AOM and SR rates reached approximately 0.3 mmol $g_{VSS}^{-1}$ $d^{-1}$. These rates are comparable or even higher than the *in vitro* AOM rates of ANME-1 or ANME-2 dominated biomass, e.g. the rate obtained after the enrichment of Eckernförde Bay sediment dominated by ANME-2 type cells for more than 800 days in a continuous membrane bioreactor (Meulepas et al., 2009a). Moreover, the AOM-SR rate measured in this study at 0.45 MPa is even higher than the AOM rate coupled to denitrification, which is thermodynamically more favorable ($\Delta G^{0'}$ = -924 kJ $mol^{-1}$ $CH_4$) (Deutzmann & Schink, 2011) than AOM-SR (Eq. 4.1).

It should be noted that even after two months incubation, the abundance of the responsible microorganisms, i.e. all detected types of ANME and SRB cells, is quite low: $17.8 \times 10^5$ and $11.4 \times 10^5$ number of copies per mL of wet sediment of ANME-3 and ANME-2, respectively in the total community (Appendix 1 Tables S4.1 and S4.2). The ANME-3 cells present in the marine Lake Grevelingen posses high specific AOM-SR rate and thus, can be of great potential to be applied in the industry after enrichment. The SR rate with $CH_4$ as electron donor should be around 100 mmol $g_{VSS}^{-1}$ $d^{-1}$ to be competitive with the rates achieved with other electron donors, such as hydrogen or ethanol (Suarez-Zuluaga et al., 2014; Meulepas et al., 2009a), which is still much higher than what was obtained in this study.

Methanogenic activity in marine Lake Grevelingen sediment was previously described by Egger et al. (2016) and confirmed in this study at low pressure (0.1 MPa) or when no $CH_4$ was added (Appendix 1, Figure S4.2). At 0.1 MPa, the $CH_4$ production rate was 31 µmol $g_{VSS}^{-1}$ $d^{-1}$ and the AOM rate was 186 µmol $g_{VSS}^{-1}$ $d^{-1}$. Trace $CH_4$ oxidation occurs during methanogenesis and the archaea involved compete with SRB for carbon sources (Meulepas et al., 2010b;

Timmers et al., 2015b). Thus, the determined AOM at 0.1 MPa cannot account for the net AOM-SR.

At high pressures (0.45, 10 MPa), AOM-SR was preferred (Figures 4.2a and 4.4a) over methanogenesis (Appendix 1, Figure S4.2). Methanogenesis becomes less thermodynamically favorable at high pressures, from 0.1 to 10 MPa, 12 kJ mol$^{-1}$ less of free energy is released (Meulepas et al., 2010b). Timmers et al. (2015b) found that at 10 MPa net AOM-SR occurred, while at 0.1 MPa methanogenesis and trace $CH_4$ oxidation dominated. In this study, the optimal AOM-SR was 0.45 MPa: the SR activity decreased at pressures higher than 10 MPa, while AOM activity already decreased at pressures higher than 0.45 MPa (Figures 4.2a and 4.4a).

## 4.5 Conclusions

This is the first study showing that the highest AOM-SR activity of a marine sediment, sampled from a shallow marine lake and predominantly containing ANME-3, occurs at low pressures (0.1 and 0.45 MPa). The active ANME adapted to coastal marine Lake Grevlingen sediment preferred a lower $CH_4$ concentration over elevated pressures (10, 20, 40 MPa), in contrast to previous studies that show strong positive correlations between the growth of ANME-1/2 and the $CH_4$ pressure. Pressure steered the abundance and structure of the different types of ANME and SRB. The ANME-3 type was predominantly enriched in incubations at low pressures, whereas high pressures enhanced ANME-2 proliferation. Similarly, a shift from sulfate reducers to sulfur reducers was observed in the bacterial community from low (0.1 and 0.45 MPa) to high (10, 20, 40 MPa) $CH_4$ partial pressure. This research highlights that ANME-3 from marine Lake Grevelingen can be enriched at low rather than high $CH_4$ partial pressure, which is important to further understand their metabolism and physiology.

## 4.6 References

APHA. 1995. Standard methods for the examination of water and wastewater. *American Public Health Association*, (19th edition), Washington DC, USA. pp.1325.

Bhattarai, S., Cassarini, C., Gonzalez-Gil, G., Egger, M., Slomp, C.P., Zhang, Y., Esposito, G., Lens, P.N.L. 2017. Anaerobic methane-oxidizing microbial community in a coastal marine sediment: anaerobic methanotrophy dominated by ANME-3. *Microb. Ecol.* **74**(3), 608-622.

Boetius, A., Ravenschlag, K., Schubert, C.J., Rickert, D., Widdel, F., Gieseke, A., Amann, R., Jørgensen, B.B., Witte, U., Pfannkuche, O. 2000. A marine microbial consortium apparently mediating anaerobic oxidation of methane. *Nature*, **407**(6804), 623-626.

Daims, H., Brühl, A., Amann, R., Schleifer, K.-H., Wagner, M. 1999. The domain-specific probe EUB338 is insufficient for the detection of all Bacteria: development and evaluation of a more comprehensive probe set. *Syst. Appl. Microbiol.,* **22**(3), 434-444.

Daly, K., Sharp, R.J., McCarthy, A.J. 2000. Development of oligonucleotide probes and PCR primers for detecting phylogenetic subgroups of sulfate-reducing bacteria. *Microbiology,* **146**(7),1693-1705.

Deusner, C., Meyer, V., Ferdelman, T. 2009. High-pressure systems for gas-phase free continuous incubation of enriched marine microbial communities performing anaerobic oxidation of methane. *Biotechnol. Bioeng.* **105**(3), 524-533.

Deutzmann, J.S., Schink, B. 2011. Anaerobic oxidation of methane in sediments of Lake Constance, an oligotrophic freshwater lake. *Appl. Environ. Microbiol.*, **77**(13), 4429-4436.

Duan, Z and S Mao (2006). A thermodynamic model for calculating methane solubility, density and gas phase composition of methane-bearing aqueous fluids from 273 to 523K and from 1 to 2000bar. *Geochimic. Cosmochim. Ac.*, **70**(13), 3369-3386.

Egger, M., Lenstra, W., Jong, D., Meysman, F.J., Sapart, C.J., van der Veen, C., Röckmann, T., Gonzalez, S., Slomp, C.P. 2016. Rapid sediment sccumulation results in high methane effluxes from coastal sediments. *PloS ONE*, **11**(8), e0161609.

Gonzalez-Gil, G., Meulepas, R.J.W., Lens, P.N.L. 2011. Biotechnological aspects of the use of methane as electron donor for sulfate reduction. in: Murray, M.-Y. (Ed.), *Comprehensive Biotechnology*, Vol. 6 (2$^{nd}$ edition), Elsevier B.V. Amsterdam, the Netherlands, pp. 419-434.

Hagens, M., Slomp, C., Meysman, F., Seitaj, D., Harlay, J., Borges, A., Middelburg, J. 2015. Biogeochemical processes and buffering capacity concurrently affect acidification in a seasonally hypoxic coastal marine basin. *Biogeosciences*, **12**(5), 1561-1583.

Hinrichs, K.-U., Hayes, J.M., Sylva, S.P., Brewer, P.G., DeLong, E.F. 1999. Methane-consuming archaebacteria in marine sediments. *Nature*, **398**(6730), 802-805.

Kamyshny, A., Borkenstein C.G., Ferdelman, T.G. 2009. Protocol for quantitative detection of elemental sulfur and polysulfide zero-valent sulfur distribution in natural aquatic samples. *Geostand. Geoanal. Res.,* **33**(3), 415-435.

Kamyshny, A., Ekeltchik, I., Gun J., Lev, O. 2006. Method for the determination of inorganic polysulfide distribution in aquatic systems. *Anal. Chem.,* **78**(8), 2631-2639.

Kamyshny, A., Goifman, A., Gun, J., Rizkov, D., Lev, O. 2004. Equilibrium distribution of polysulfide ions in aqueous solutions at 25 °C: a new approach for the study of polysulfides' equilibria. *Environ. Sci. Technol.*, **38**(24), 6633-6644.

Knittel, K., Lösekann, T., Boetius, A., Kort, R., Amann, R. 2005. Diversity and distribution of methanotrophic archaea at cold seeps. *Appl. Environ. Microbiol.*, **71**(1), 467-479.

Krüger, M., Treude, T., Wolters, H., Nauhaus, K., Boetius, A. 2005. Microbial methane turnover in different marine habitats. *Palaeogeogr. Palaeocl.*, **227**(1-3), 6-17.

Liamleam, W., Annachhatre, A.P. 2007. Electron donors for biological sulfate reduction. *Biotechnol. Adv.*, **25**(5), 452-463.

Losekann, T., Knittel, K., Nadalig, T., Fuchs, B., Niemann, H., Boetius, A., Amann, R. 2007. Diversity and abundance of aerobic and anaerobic methane oxidizers at the Haakon Mosby mud volcano, Barents Sea. *Appl. Environ. Microbiol.*, **73**(10), 3348-3362.

McGlynn, S.E., Chadwick, G.L., Kempes, C.P., Orphan, V.J. 2015. Single cell activity reveals direct electron transfer in methanotrophic consortia. *Nature*, **526**(7574), 531-535.

Meulepas, R.J.W., Jagersma, C.G., Gieteling, J., Buisman, C.J.N., Stams, A.J.M., Lens, P.N.L. 2009a. Enrichment of anaerobic methanotrophs in sulfate-reducing membrane bioreactors. *Biotechnol. Bioeng.*, **104**(3), 458-470.

Meulepas, R.J.W., Jagersma, C.G., Khadem, A.F., Buisman, C.J.N., Stams, A.J.M., Lens, P.N.L. 2009b. Effect of environmental conditions on sulfate reduction with methane as electron donor by an Eckernförde Bay enrichment. *Environ. Sci. Technol.*, **43**(17), 6553-6559.

Meulepas, R.J.W., Stams, A.J.M., Lens, P.N.L. 2010a. Biotechnological aspects of sulfate reduction with methane as electron donor. *Rev. Environ. Sci. Biotechnol.*, **9**(1), 59-78.

Meulepas, R.J.W., Jagersma, C.G., Zhang, Y., Petrillo, M., Cai, H., Buisman, C.J.N., Stams, A.J.W., Lens, P.N.L. 2010b. Trace methane oxidation and the methane dependency of sulfate reduction in anaerobic granular sludge. *FEMS Microbiol. Ecol.*, **72**(2), 261-271.

Milucka, J., Ferdelman, T.G., Polerecky, L., Franzke, D., Wegener, G., Schmid, M., Lieberwirth, I., Wagner, M., Widdel, F., Kuypers, M.M.M. 2012. Zero-valent sulphur is a key intermediate in marine methane oxidation. *Nature*, **491**(7425), 541-546.

Nauhaus, K., Boetius, A., Kruger, M., Widdel, F. 2002. *In vitro* demonstration of anaerobic oxidation of methane coupled to sulphate reduction in sediment from a marine gas hydrate area. *Environ. Microbiol.*, **4**(5), 296-305.

Nadkarni, M.A., Martin, F.E., Jacques, N.A., and Hunter, N. (2002) Determination of bacterial load by real-time PCR using a broad-range (universal) probe and primers set. *Microbiol-Sgm*, **148**(Pt 1), 257-266.

Niemann, H., Losekann, T., de Beer, D., Elvert, M., Nadalig, T., Knittel, K., Amann, R., Sauter, E.J., Schluter, M., Klages, M., Foucher, J.P., Boetius, A. 2006. Novel microbial communities of the Haakon Mosby mud volcano and their role as a methane sink. *Nature*, **443**(7113), 854-858.

Omoregie, E.O., Mastalerz, V., de Lange, G., Straub, K.L., Kappler, A., Røy, H., Stadnitskaia, A., Foucher, J.-P., Boetius, A. 2008. Biogeochemistry and community composition of iron- and sulfur-precipitating microbial mats at the Chefren mud volcano (Nile Deep Sea Fan, Eastern Mediterranean). *Appl. Environ. Microbiol.*, **74**(10), 3198-3215.

Reeburgh, W.S. 2007. Oceanic methane biogeochemistry. *Chem. Rev.*, **107**(2), 486-513.

Scheller, S., Yu, H., Chadwick, G.L., McGlynn, S.E., Orphan, V.J. 2016. Artificial electron acceptors decouple archaeal methane oxidation from sulfate reduction. *Science*, **351**(6274), 703-707.

Shigematsu, T., Tang, Y., Kobayashi,T., Kawaguchi, H., Morimura S., Kida, K. 2004. Effect of dilution rate on metabolic pathway shift between aceticlastic and nonaceticlastic methanogenesis in chemostat cultivation. *Appl. Environ. Microbiol.*, **70**(7), 4048-4052.

Sievert, S.M., Kiene, R.P., Schulz-Vogt, H.N. 2007. The sulfur cycle. *Oceanography*, **20**(2), 117-123.

Snaidr, J., Amann, R., Huber, I., Ludwig, W., Schleifer, K.H. 1997. Phylogenetic analysis and *in situ* identification of bacteria in activated sludge. *Appl. Environ. Microbiol.*, **63**(7), 2884-2896.

Song, Z.-Q., Wang, F.-P., Zhi, X.-Y., Chen, J.-Q., Zhou, E.-M., Liang, F., Xiao, X., Tang, S.-K., Jiang, H.-C., Zhang, C.L., Dong, H., Li, W.-J. 2013. Bacterial and archaeal diversities in Yunnan and Tibetan hot springs, China. *Environ. Microbiol.*, **15**(4), 1160-1175.

Stahl, D.A., Amann, R.I. 1991. Development and application of nucleic acid probes. In: Stackebrandt, E., Goodfellow, M. (Eds.), *Nucleic acid techniques in bacterial systematics.* John Wiley & Sons Ltd, Chichester, UK, pp. 205-248.

Suarez-Zuluaga, D.A., Timmers, P.H.A., Plugge, C.M., Stams, A.J.M., Buisman, C.J.N., Weijma, J. 2015. Thiosulphate conversion in a methane and acetate fed membrane bioreactor. *Environ. Sci. Pollut. R.* **23**(3), 2467-2478.

Sulu-Gambari, F., Seitaj, D., Meysman, F.J., Schauer, R., Polerecky, L., Slomp, C.P. 2016. Cable bacteria control iron-phosphorus dynamics in sediments of a coastal hypoxic basin. *Environ. Sci. Technol.* **50**(3), 1227-1233.

Takai, K., Horikoshi, K.. 2000. Rapid detection and quantification of members of the archaeal community by quantitative PCR using fluorogenic probes. *Appl. Environ. Microbiol.*, **66**(11), 5066-5072.

Timmers, P.H., Suarez-Zuluaga, D.A., van Rossem, M., Diender, M., Stams, A.J., Plugge, C.M. 2015a. Anaerobic oxidation of methane associated with sulfate reduction in a natural freshwater gas source. *ISME J.*, **10**(6), 1400-1412.

Timmers, P.H., Gieteling, J., Widjaja-Greefkes, H.A., Plugge, C.M., Stams, A.J., Lens, P.N.L., Meulepas, R.J. 2015b. Growth of anaerobic methane-oxidizing archaea and sulfate-reducing bacteria in a high-pressure membrane capsule bioreactor. *Appl. Environ. Microbiol.*, **81**(4), 1286-1296.

Treude, T., Knittel, K., Blumenberg, M., Seifert, R., Boetius, A. 2005. Subsurface microbial methanotrophic mats in the Black Sea. *Appl. Environ. Microbiol.*, **71**(10), 6375-6378.

Vigneron, A., Cruaud, P., Pignet, P., Caprais, J.-C., Cambon-Bonavita, M.-A., Godfroy, A., Toffin, L. 2013. Archaeal and anaerobic methane oxidizer communities in the Sonora Margin cold seeps, Guaymas Basin (Gulf of California). *ISME J.*, **7**(8), 1595-1608.

Wagner, M., Amann, R., Lemmer, H., Schleifer, K.-H. 1993. Probing activated sludge with oligonucleotides specific for proteobacteria: inadequacy of culture-dependent methods for describing microbial community structure. *Appl. Environ. Microbiol.*, **59**(5), 1520-1525.

Wegener, G., Krukenberg, V., Ruff, S.E., Kellermann, M.Y., Knittel, K. 2016. Metabolic capabilities of microorganisms involved in and associated with the anaerobic oxidation of methane. *Front. Microbiol.*, **7**(46), 1-16.

Yamamoto, S., Alcauskas, J.B., Crozier, T.E. 1976. Solubility of methane in distilled water and seawater. *J. Chem. Eng. Data*, **21**(1), 78-80.

Zhang, Y., Henriet, J.-P., Bursens, J., Boon, N. 2010. Stimulation of in vitro anaerobic oxidation of methane rate in a continuous high-pressure bioreactor. *Biores. Technol.*, **101**(9), 3132-3138.

Zhang, Y., Li, H., Bartlett D.H., Xiao, X. 2015. Current developments in marine microbiology: high-pressure biotechnology and the genetic engineering of piezophiles. *Curr. Opin. Microbiol.*, **33**(13), 157-164.

Zhang, Y., Maignien, L., Stadnitskaia, A., Boeckx, P., Xiao, X. Boon, N. 2014. Stratified community responses to methane and sulfate supplies in mud volcano deposits: insights from an *in vitro* experiment. *PLoS ONE*, **9**(11), 1-9.

# CHAPTER 5

## Anaerobic Oxidation of Methane Coupled to Thiosulfate Reduction in a Biotrickling Filter

This chapter has been published as:

Cassarini C., Rene E. R., Bhattarai S., Esposito G., Lens P.N.L. (2017) Anaerobic oxidation of methane coupled to thisoulfate reduction in a biotrickling filter. *Bioresour. Technol.*, **240**(3), 214-222.

**Abstract**

Microorganisms from an anaerobic methane oxidizing sediment were enriched with methane gas as the substrate in a biotrickling filter (BTF) using thiosulfate as electron acceptor for 213 days. Thiosulfate disproportionation to sulfate and sulfide was the dominating sulfur conversion process in the BTF, the sulfide production rate was 0.5 mmol $l^{-1}$ $day^{-1}$. A specific group of sulfate reducing bacteria (SRB), belonging to the *Desulforsarcina/Desulfococcus* group, was enriched in the BTF. The BTF biomass had a maximum sulfate reduction rate with methane as sole electron donor of 0.38 mmol $l^{-1}$ $day^{-1}$, measured in the absence of thiosulfate in the BTF. Therefore, a BTF fed with thiosulfate as electron acceptor can be used to enrich SRB of the DSS group and activate the inoculum for anaerobic oxidation of methane coupled to sulfate reduction.

## 5.1 Introduction

Sulfate and other sulfur oxyanions, such as thiosulfate, sulfite or dithionite, are contaminants discharged in fresh water due to industrial activities such as food processing, fermentation, coal mining, tannery and paper processing. Biological treatment of these wastewaters has been successfully applied wherein the sulfur oxyanions are anaerobically reduced to sulfide, which is then either oxidized to elemental sulfur or precipitated as metal sulfide (Liamleam and Annachhatre, 2007; Weijma et al., 2006). Many sulfate rich wastewaters are deficient in electron donor and the addition of an external carbon source is often required to achieve complete sulfate reduction by sulfate reducing bacteria (SRB). Electron donors such as ethanol, methanol, hydrogen, acetate, lactate and propionate are usually supplied, but these increase the operational and investment costs (Meulepas et al., 2010). Therefore, the use of easily accessible and low-priced electron donors such as methane is appealing (Gonzalez-Gil et al., 2011). Moreover, methane is also a well known green house gas and its increase in atmospheric concentration could have large implications for future climate change (Forster et al., 2007). Besides, the surface layers of wetlands, sediments, paddy fields and several other terrestrial and aquatic surfaces are known to produce methane and hence, reducing its concentration in the atmosphere is thus important (Kirschke et al., 2013).

Anaerobic oxidation of methane (AOM) coupled to sulfate reduction is a naturally occurring process in anaerobic environments, such as in marine sediments. This process is mediated by a special group of slow growing and so far uncultured anaerobic methanotrophs (ANME) and SRB that can thrive in harsh environments by using the abundance of methane and $H_2S$ present in such habitats. ANME are grouped into three distinct clades, i.e. ANME-1, ANME-2 and ANME-3. The common SRB associated with ANME are *Desulfosarcina/Desulfococcus* (DSS) and *Desulfobubaceae* (Schreiber et al., 2010).

The main challenge of using AOM coupled to sulfate reduction (AOM-SR) as a process for methane removal and desulfurization of wastewater is the slow growth rate of the microorganisms involved (Meulepas et al., 2009; Zhang et al., 2010). The highest AOM-SR rates reported so far in the literature (0.6 mmol $l^{-1}$ $day^{-1}$ (Meulepas et al., 2009) are too low (~100 times lower) to economically compete with the electron donors hydrogen or ethanol

(Meulepas et al., 2009; Suarez-Zuluaga et al., 2014). The AOM-SR rates could be increased by using more thermodynamically favorable sulfur compounds, such as thiosulfate (Eq. 5.1) or by growing them in a bioreactor with high biomass retention capability, such as membrane bioreactors (Meulepas et al., 2009).

$$CH_4 + S_2O_3^{2-} \rightarrow HCO_3^- + 2HS^- + H^+ \qquad \Delta G^{0'} = -39 \text{ kJ mol}^{-1} \text{ CH}_4 \qquad \text{Eq. 5.1}$$

In this study, a commonly used wastegas treatment but so far not in AOM studies, the reactor type biotrickling filter (BTF), was used to enrich the microorganisms involved in the AOM coupled to thiosulfate reduction and to increase the rates of sulfide production and methane oxidation. The inoculum used was collected from an active AOM site (Alpha Mound, Gulf of Cadiz). However, the *in situ* or *ex situ* AOM-SR rate of the Alpha Mound sediment has not yet been estimated and the specific group of microorganisms involved has not yet been investigated.

The polyurethane foam was used as the packing material of the BTF because of its high porosity, good biomass retention capacity and its ability to enhance gas to liquid mass transfer of the poorly soluble methane by increasing gas-liquid mixing and retaining methane in the pores (Aoki et al., 2014; Estrada et al., 2014). The carbon and sulfur bioconversions of the consortia growing on the polyurethane foam and the possible abiotic processes were assessed with the help of batch tests and the microorganisms enriched in the BTF after long term operation (213 days) were visualized and identified.

## 5.2    Material and methods

### 5.2.1    Source of sediment biomass

Sediment samples were obtained from the Alpha Mound (35°17.48'N; 6°47.05'W, water depth ca. 528 m), Gulf of Cadiz (Spain), during R/V Marion Dufresne Cruise MD 169 MICROSYSTEMS to the Gulf of Cadiz in July 2008. The Gulf of Cadiz is located in the eastern Atlantic ocean, North West of the Strait of Gibraltar, along the Spanish and Portuguese continental margin (Niemann et al., 2006). This is an area of mud volcanism and gas venting through the seafloor. Moreover, cold-water coral carbonate mounds, such as the Alpha Mound, have been discovered at the Pen Duick escarpment on the Moroccan margin (Maignien et al., 2010). In previous studies, the Alpha Mound showed evidence for the presence of a shallow sulfate-methane transition zone at ~300 cm sediment depth with increased sulfate reduction rates indicating the presence of microbial mediated AOM (Templer et al., 2011).

Sediment samples were recovered by gravity coring from Alpha Mound, retrieving up to 4.3 m of sediment. Gravity cores were sectioned into 1 m sections and immediately stored at 4 °C. The cores were then opened, subsampled (the sampling interval for all parameters was 10 to 20 cm) (Templer et al., 2011; Wehrmann et al., 2011), capped and stored at 4 °C in trilaminate polyetherimide coated aluminum bags (KENOSHA C.V., Amstelveen, the Netherlands) under nitrogen rich atmosphere (Zhang et al., 2010). The sediment used in this study was retrieved from 250 to 270 cm below the sea floor and was stored at 4 °C with a headspace of methane for five years before it was inoculated into the BTF.

### 5.2.2 Composition of the artificial seawater medium

The artificial seawater based liquid medium used in the BTF had the following composition per liter of demineralised water (Zhang et al., 2010): NaCl (26 g), KCl (0.5 g), $MgCl_2 \cdot 6H_2O$ (5 g), $NH_4Cl$ (0.3 g), $CaCl_2 \cdot 2H_2O$ (1.4 g), $Na_2S_2O_3$ (1.6 g), $KH_2PO_4$ (0.1 g), trace element solution (1 ml), 1 M $NaHCO_3$ (30 ml), vitamin solution (1 ml), thiamin solution (1 ml), vitamin $B_{12}$ solution (1 ml), 0.5 g $L^{-1}$ resazurin solution as a redox indicator (1 ml) and 0.5 M $Na_2S$ solution (1 ml). The vitamins and trace element solution were prepared according to the protocol outlined by Widdel and Bak (1992). The pH was adjusted to 7.0 with sterile 1 M $Na_2CO_3$ or $H_2SO_4$ solutions, which was stored under nitrogen atmosphere. All chemicals were purchased as lab grade in anhydrous form from Fisher Scientific (Sheepsbouwersweg, the Netherlands). The medium was kept anoxic with the help of nitrogen purging until it was recirculated within the BTF.

### 5.2.3 BTF setup and operation

The BTF (Figure 5.1) consisted of an acrylic pipe (height 32 cm and diameter 55 mm), sealed air-tight to prevent leakage or air intrusion during its operation. The filter bed volume of the reactor was 0.4 l, which was packed with polyurethane foam cubes of 1 $cm^3$ (98% porosity and a density of 28 kg $m^{-3}$) and 20 ml of the sampled Alpha Mound sediment ($0.03 \pm 0.01$ g volatile suspended solids). Two circular acrylic sieve plates (pore size of 3.5 mm) were placed at the bottom and top of the BTF to hold the polyurethane foam pieces (Figure 5.1).

The BTF was operated in sequential fed-batch mode for the influent (artificial seawater), while the methane gas (99.5% methane, Linde gas, Schiedam, the Netherlands) stored in Tedlar bags was continuously supplied to the bioreactor using a peristaltic pump (Verder International BV, Utrecht, the Netherlands) at a flow rate of 2 ml $min^{-1}$. The estimated empty bed residence time of methane was 200 min. The BTF was operated in a counter-current mode: the gas was passed from the bottom of the BTF to the top, while the seawater medium was recirculated from the top to the bottom. The medium trickled uniformly over the entire cross sectional area of the packing through a spray head having a pore size of 4.0 mm. The trickled medium flowed into the nutrient holding tank (1.5 l), which was then continuously recirculated to the BTF with the help of a Masterflex S/L peristaltic pump (Metrohm Netherlands B.V., Schiedam, the Netherlands) operating at a flow rate of 10 ml $min^{-1}$ (Figure 5.1).

The BTF was operated for 213 days and it was maintained in the dark and at room temperature ($\sim 20 \pm 2$ °C). During BTF operation, the seawater medium containing 10 mM thiosulfate was replaced periodically (days 38, 104, 139, 189 and 206, Figure 5.2), on day 91 thiosulfate was added to the not refreshed medium (Figure 5.2) and from days 46 to 88 the BTF was operated in the absence of thiosulfate (Figure 5.2). Both gas (in and out) and BTF effluent were sampled twice a week from the sampling ports (Figure 5.1). pH, sulfate, sulfide and thiosulfate concentrations were measured in samples collected from the liquid medium, while methane and carbon dioxide were analyzed from the gas samples collected at the inlet and outlet of the BTF (Figure 5.1). Biomass samples for microbial visualization were obtained before inoculation and at the end of the BTF operation (day 213).

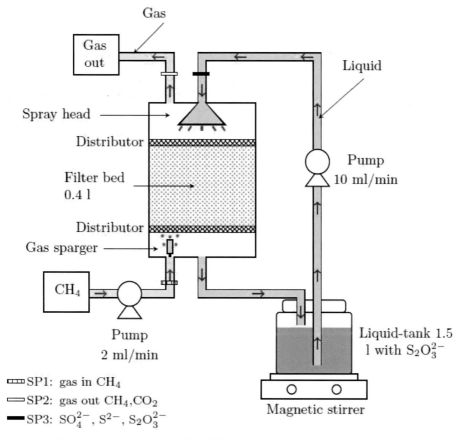

**Figure 5.1** Schematic of a biotrickling filter (BTF) configuration for anaerobic methane oxidation coupled to thiosulfate reduction.

### 5.2.4  Abiotic batch tests

Abiotic disproportionation, the effect of polyurethane foam and the possible formation of polysulfides under abiotic conditions were assessed by performing batch tests. The original Alpha Mound sediment was homogenized in an anaerobic chamber and diluted with the artificial seawater medium in a 1:30 ratio, and it was aliquoted in 120 ml sterile serum bottles (40 ml headspace with methane). Duplicates were prepared for each type of incubation as follows: presence of polyurethane foam pieces and sediment, polyurethane foam pieces only, sediment only, and absence of both polyurethane foam pieces and sediment. The bottles were closed with butyl rubber stoppers, sealed with aluminum crimp and flushed with methane (99.5%, Linde gas, Schiedam, the Netherlands) to 0.2 MPa of pressure. The bottles were incubated in the dark at ~ 20 ($\pm$ 2) °C. The batch tests were performed for 63 days and sampling was done eight times for sulfide, thiosulfate and sulfide by withdrawing 500 µl of liquid sample. Headspace analysis for methane and carbon dioxide was done only twice by withdrawing 500 µl of gas sample. The gas samples were measured in duplicate along with controls for quality assurance.

### 5.2.5 Chemical analysis

The pH was measured with a Metrohm pH meter (Metrohm Applikon B.V., Schiedam, the Netherlands) and a pH electrode (SenTix WTW, Amsterdam, the Netherlands). Sulfate and thiosulfate were analyzed using an Ion Chromatograph system (Dionex-ICS-1000 with AS-DV sampler) as described previously (Villa-Gomez et al., 2011). Total dissolved sulfide concentrations were analyzed spectrophotochemically using the methylene blue method (Acree et al., 1971) and the amount of sulfide measured accounted for all cumulative dissolved sulfide species ($H_2S$, $HS^-$ and $S^{2-}$) in the BTF. Duplicate measurements were done for the analysis of pH, sulfate, thiosulfate and dissolved sulfide to evaluate the standard deviation.

Methane and carbon dioxide concentrations from the inlet and outlet of the BTF and from the headspace of the batches were measured by injecting 0.5 ml sample in a gas chromatograph (GC 3800, VARIAN, Middelburg, the Netherlands). The gas chromatograph was equipped with a PORABOND Q column (25 m × 0.53 mm × 10 μm) and equipped with a thermal conductivity detector. The carrier gas was helium (15 Psi) at a flow rate of 0.5 ml min$^{-1}$, while the oven temperature was maintained at 25 °C. For each sampling, gas measurements were performed in duplicates and the data used for the analysis had a standard deviation lower than 0.5%. Standard gas mixtures of methane and carbon dioxide were measured every time along with sample measurements.

### 5.2.6 Biological analysis

Microbial analysis of biomass collected from the BTF was performed by catalyzed reporter deposition-fluorescence *in situ* hybridization (CARD-FISH). Sediment samples were fixed in 1% paraformaldehyde/phosphate buffer saline (PBS) (*v/v*) overnight at 4 °C, then washed with PBS and stored in 50% ethanol/PBS (*v/v*) at -20 °C until it was processed further. The fixed samples were filtered on polycarbonate filters and embedded in low-gelling point agarose (Pernthaler et al., 2002; Wendeberg, 2010). For bacterial cell wall permeabilization, samples were treated with 10 mg ml$^{-1}$ of lysozyme solution and subsequently with 60 U ml$^{-1}$ achromopeptidase solution (Sekar et al., 2003; Wendeberg, 2010). For archaeal cell wall permeabilization, filters were incubated with a sodium dodecyl sulfate and proteinase K solution as described by Holler et al. (2011). For all samples, endogenous peroxidases were inactivated with 0.1% $H_2O_2$ as described by Wendeberg (2010).

CARD-FISH with horseradish peroxidase (HRP)-labeled oligonucleotide probes and tyramide signal amplification was done according to previously described protocols (Pernthaler et al., 2002; Pernthaler et al., 2004), using the fluorochrome Oregon Green 488-X (Molecular Probes, Eugene, OR). The microorganisms were visualized using archaeal and bacterial HRP-labeled oligonucleotide probes ARCH915 (Stahl and Amann, 1991) and EUB338-I-III (Daims et al., 1999), respectively. The probes DSS658 (Manz et al., 1998) and ANME-2 538 (Schreiber et al., 2010) were used for the detection of DSS and ANME-2, respectively. Oligonucleotide probes were purchased from Biomers (Ulm, Germany). Finally, all the cells were stained with 4', 6'-diamidino-2-phenylindole (DAPI) and analyzed using an epifluorescence microscope (Carl Zeiss, Germany).

## 5.3    Results and discussion

### 5.3.1    Performance of the BTF

Methane was the sole electron donor available for the microorganisms present in the BTF. Hypothetically, methane should be oxidized to bicarbonate, while one mole of thiosulfate should be reduced to two moles of sulfide, following Eq. 5.1. However, during phase I of BTF operation (Figure 5.2b), thiosulfate was consumed from the beginning while sulfate was produced (Figure 5.2b), suggesting that microbial disproportionation of thiosulfate to sulfide and sulfate occurred in the BTF (Eq. 5.2).

$$S_2O_3^{2-} + H_2O \rightarrow SO_4^{2-} + HS^- + H^+ \qquad\qquad \text{Eq. 5.2}$$

In phase I, the decrease of the thiosulfate concentration was in a ratio of -1:+0.96 to the increase of the sulfate concentration, according to Eq. 5.2. Thiosulfate consumption and sulfate production rates were 0.39 and 0.31 mmol $l^{-1}$ day$^{-1}$, respectively. However, the amount of sulfate produced was higher than the amount of total dissolved sulfide produced in the BTF. At the end of phase I, the concentration of sulfur removed as thiosulfate was 19.5 mM, but only 9.4 mM (48%) was recovered as dissolved sulfide and sulfate. Chemical oxidation of sulfide to elemental sulfur or sulfate might have occurred by iron oxides (Thamdrup et al., 1994), already present in the sediment (300 mg $l^{-1}$ of total iron was measured in the sediment before the incubation). Around 6 mM of reactive iron oxides were needed to oxidize the produced sulfide to elemental sulfur. In the original Alpha Mound sediment, which was used as inoculum for the BTF, reducible iron-(oxyhydr)oxides are present in the top layer of the sediment and high concentrations of pyrite were detected concomitant to the sulfate reduction zone (Wehrmann et al., 2011). Therefore, as previously described in other studies (Finster et al., 1998; Wan et al., 2014), it is possible that the sediment inherently contained enough reactive iron that could have oxidized the sulfide to elemental sulfur or precipitated it as pyrite (FeS$_2$), and was thus not detected by the method used to determine the sulfide concentration.

When the total dissolved sulfide concentration reached 4.4 mM, the medium in the nutrient tank (Figure 5.1) changed color from transparent to yellow, after which the total dissolved sulfide concentration decreased (phase I, Figure 5.2b). The yellow coloration suggests the formation of soluble polysulfides (Kamyshny et al., 2007) according to Eq. 5.3:

$$(n-1)S^0 + HS^- \rightarrow S_n^{2-} + H^+ \qquad\qquad \text{Eq. 5.3}$$

The produced elemental sulfur, although not quantified in this study, can form polysulfides with different numbers of sulfur atoms in the presence of sulfide at alkaline pH (Kamyshny et al., 2007; Poser et al., 2013). This hypothesis was confirmed by the change in pH (from 8.0 to 7.2) of the liquid medium before and after the color change of the medium (Figure 5.2a). In the study of Cypionka et al. (1998), the pathway of thiosulfate disproportionation was investigated and they proposed the formation of elemental sulfur and sulfite as intermediates followed by their disproportionation to sulfide and sulfate. The latter pathway could explain the delayed sulfide production in the BTF (Figure 5.2b); however, further studies on identifying the thiosulfate metabolic pathway by the studied Alpha Mound sediment are necessary (Finster, 2008). In phase I, the methane consumption and carbon dioxide production were high (Figure

5.2c) before and after the complete change of color of the medium.

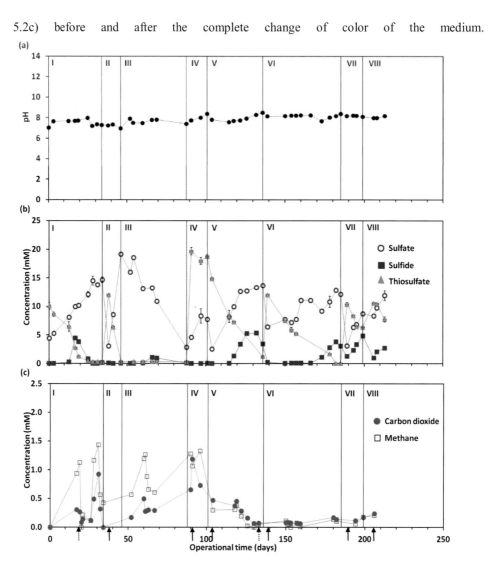

**Figure 5.2** Profiles of different process parameters monitored during the operation of the BTF with methane as an electron donor and thiosulfate as electron acceptor: (a) pH; (b) thiosulfate, sulfate and sulfide; and (c) methane and carbon dioxide. The vertical lines represent the different phases in bioreactor operation; I: days 0-34, II: days 34-46, III: days 46-88, IV: days 88-101, V: days 101-136, VI: days 136-185, VII: days 185-213. The black arrows indicate the days at which the mineral medium was replaced, while the dashed arrows indicate the days at which yellow coloration of the mineral medium was observed. Error bars represent the standard deviation of duplicate measurements.

In phase II, the mineral medium was replenished and thiosulfate was depleted within 8 days with hardly any sulfide production (less than 0.01 mmol l⁻¹ day⁻¹); however, sulfate was produced at the highest rate (2 mmol l⁻¹ day⁻¹) during this phase of BTF operation. In phase III, no thiosulfate and sulfate were added to the influent to ascertain whether the newly formed sulfate from the thiosulfate disproportionation in phase III could be reduced concomitant to the oxidation of methane. The sulfate reduction rate during phase III was 0.38 mmol l⁻¹ day⁻¹

(Figure 5.2b) and the profiles of methane and carbon dioxide suggested the oxidation of methane (Figure 5.2c). Therefore, the produced sulfate from thiosulfate disproportionation in phase II seemed to be readily available for the SRB to be reduced in phase III (Figure 5.2b). The sulfate reduction rate (0.38 mmol $l^{-1}$ day$^{-1}$) achieved in phase III was lower than the highest volumetric rate reported by Meulepas et al. (2009) in a membrane bioreactor (0.6 mmol $l^{-1}$ day$^{-1}$) inoculated with Eckernförde Bay sediment. This rate was obtained after 884 days of BTF operation of the bioreactor and it required a long start-up period of up to 400 days. Long start-up periods (~365 days) have also been reported in continuous-flow bioreactors with polyurethane sponges, incubated with deep sea methane-seep sediments collected from the Nankai Trough, Japan (Aoki et al., 2014). Differently, in this study the sulfate reduction occurred immediately after thiosulfate was completely depleted (Phase III, Figure 5.2b). However, hardly any total dissolved sulfide was detected until day 66, wherein the concentration was only 1 mM (Phase III, Figure 5.2b). This suggested that an unknown sulfur compound might have been formed, even though no change of color of the medium was observed. Presumably, in this case, solid elemental sulfur might have accumulated in the packing material of the BTF column, as reported in the literature (Suarez-Zuluaga et al., 2015).

During phase IV, the BTF was fed with thiosulfate (Figure 5.2b) and the formation of sulfate was evidenced (0.61 mmol $l^{-1}$ day$^{-1}$), as seen previously during phases I and II. This can be attributed to disproportionation because total dissolved sulfide was not produced during this phase. In Phase V, the liquid medium was replenished with fresh medium containing thiosulfate. In this phase, the thiosulfate consumption and sulfate production rates were 0.45 and 0.49 mmol $l^{-1}$ day$^{-1}$, respectively, and sulfide was produced to a maximum concentration of 5.4 mM (0.50 mmol $l^{-1}$ day$^{-1}$), showing that thiosulfate disproportionation occurred following the stoichiometry of Eq. 5.2. In this phase, the liquid medium in the nutrient tank turned yellow as in phase I. 5.4 mM of total dissolved sulfide might have been already toxic for the microorganisms in the BTF, as 2.5 mM of dissolved sulfide was shown to be toxic for the microorganisms from Eckernförde Bay enriched in a membrane reactor (Meulepas et al., 2009) and 4 mM sulfide was shown to be toxic for high pressure incubation of sediment collected from a Mud Volcano in the Gulf of Cadiz (Zhang et al., 2011). Moreover, at day 126, before the liquid medium turned yellow, almost 70% of disproportionated sulfur from thiosulfate was recovered as dissolved sulfide and sulfate. The dissolved sulfide which was not detected could have been oxidized to elemental sulfur, as mentioned previously.

During the last three phases (phases VI, VII and VIII), the disproportionation occurred according to Eq. 5.2, and sulfide and sulfate concentrations were nearly equal (Figure 5.2b). During phases VII and VIII, 75 and 94% of thiosulfate consumed, respectively, was recovered as dissolved sulfide and sulfate. This suggests that disproportionation occurred, while less dissolved sulfide was reoxidized than in previous phases. In phases VII and VIII, the sulfate production thiosulfate consumption rates varied between 0.4 and 0.6 mmol $l^{-1}$ day$^{-1}$ as in phase V (Figure 5.2b). Differently, these values were lower in phase VI, which could be due to the high amounts of dissolved sulfide produced during phase V. The high concentration of dissolved sulfide (5.4 mM) might have been toxic for the microorganisms present in the BTF, but no complete inhibition occurred. Nevertheless, disproportionation was the dominating process in the last phases, probably without the occurrence of AOM. Thus, the SRB responsible for thiosulfate disproportionation are probably more sulfide tolerant than the AOM community and they can continue to grow at sulfide concentration over 5 mM, as previously reported (Finster et al., 1998; Poser et al., 2013).

From day 104 until the end of the BTF operation, the methane consumption and carbon dioxide profiles were nearly similar. However, when polysulfide was formed in phase V, methane consumption as well as carbon dioxide production decreased and thereafter they increased only slightly towards the end of the BTF operation (Figure 5.2c). The methane and carbon dioxide in the gas phase can indicate the possible consumption of methane, but it does not account for the formation of methane due to possible methanogenic activity or carbon dioxide production from sources other than methane. Moreover, the utilization of carbon dioxide by ANME and its bacterial partners has been investigated in other sediments and in a few studies ANME-1 (Holler et al., 2011; Treude et al., 2007) and ANME-2 and their bacterial partners (Wegener et al., 2016) have been defined as autotrophic.

In phase III, 13 mM of sulfate was reduced with a rate of 0.38 mmol $l^{-1}$ day$^{-1}$, stoichiometrically, the same amount of methane should have been oxidized. However, in the BTF operation phases in which sulfate was not reduced, methane could have also been oxidized by reactive iron oxides in the sediment. Previous studies have shown how AOM can be coupled either to sulfate or iron reduction (Egger et al., 2015) and also how some metal reducing bacteria can use either iron or sulfur as electron acceptors depending on the environmental conditions (Flynn et al., 2014). In further studies, the use of isotopic labeled methane can be used to determine the net AOM occurrence. Nevertheless, AOM likely occurred in the BTF since methane was the only electron donor available for sulfate reduction.

### 5.3.2   Effect of polyurethane foam on sulfur and carbon compound profiles

The sulfide concentration during the BTF operation was always lower than expected probably due to the precipitation with iron or the formation of elemental sulfur. Therefore, the effect of polyurethane foam pieces on the concentrations of total dissolved sulfide, thiosulfate and sulfate was tested, together with the possible occurrence of abiotic reactions. These batch tests were performed with the Alpha Mound (Gulf of Cadiz) sediment inoculum.

The results from these batch incubations showed the disproportionation of thiosulfate to sulfate and total dissolved sulfide as observed in the BTF (Figures 5.3a and 5.3c). Although sulfate was produced immediately, sulfide production started only after 24 days of incubation with and without the addition of polyurethane foam (control) (Figures 5.3a and 5.3c). Besides, the production of sulfide was less than expected from the reaction stoichiometry (Eq. 5.2). The formation of sulfate from thiosulfate was solely due to biological activity since the concentration of thiosulfate, sulfate and sulfide hardly changed during the incubation period in the batches without sediment (Figures 5.3b and 5.3d).

Dissolved sulfide concentrations reached values of 4.1 and 2.9 mM, with and without the addition of polyurethane foam pieces, respectively (Figures 5.3a and 5.3c). However, no change of color (from transparent to yellow) of the medium occurred, in contrast to the BTF medium. It is noteworthy to mention that during batch incubations, the polyurethane foam pieces were completely submerged in the seawater medium, whereas in the BTF, the medium was trickled through the foam, probably allowing the pieces to retain other products such as non soluble elemental sulfur or precipitated metal sulfide

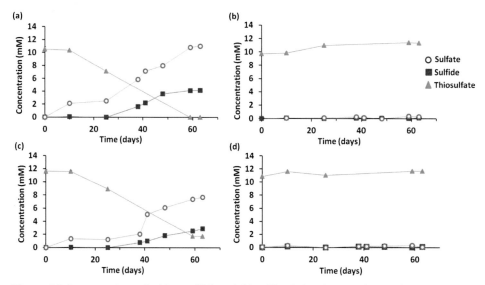

**Figure 5.3** Concentrations of sulfate, sulfide and thiosulfate in batch tests using methane as electron donor and thiosulfate as electron acceptor: (a) batch test with polyurethane foam pieces and (b) controls without sediment, (c) batch test without polyurethane foam pieces and (d) controls with killed biomass.

Similar to the dissolved sulfide profiles in the batches, more sulfate was produced in the batches with polyurethane foam pieces (4.1 mM, Figure 5.3a) than the incubations without foam pieces (2.9 mM, Figure 5.3c). Thus, the sulfide and sulfate production rates were slightly higher in the incubations with foam pieces ($0.19 \pm 0.02$ and $0.10 \pm 0.02$ mmol l$^{-1}$ day$^{-1}$, respectively) than without ($0.14 \pm 0.02$ and $0.07 \pm 0.01$ mmol l$^{-1}$ day$^{-1}$, respectively) foam pieces. The difference in rates between the incubations with and without polyurethane foam pieces was not very high (Figures 5.3a and 5.3c). Thiosulfate is a dissolved compound and thus readily available for the microorganisms, thus it could be easily disproportionated by SRB in the absence of the porous packing material. The polyurethane foam has the ability to enhance the methane to liquid mass transfer, but methane is not used during thiosulfate disproportionation (Eq. 5.2), which was the main process occurring in the batches.

The methane and carbon dioxide concentrations in the gas phase were measured only twice during the batch tests and it was observed that the methane was consumed and carbon dioxide was produced only in the incubations with the sediment (data not shown), suggesting biotic methane consumption. After 41 days, in the incubations with polyurethane foam pieces, 1 mM of methane was consumed and 1.1 mM of carbon dioxide was produced, while in the absence of the polyurethane foam pieces, 0.5 mM of methane was consumed and 0.7 mM of carbon dioxide was produced. This suggests the use of polyurethane foam could enhance the bioconversion rates, probably by facilitating the gas-solid mass transfer.

### 5.3.3 Visualization of the enriched microorganisms

In order to study the microbial communities enriched during 213 days of BTF operation, CARD-FISH was performed using oligonucleotide probes targeting the groups of SRB and ANME usually found in AOM-SR active sites. The cells retrieved at the end of the BTF operation (213 days) were stained with the general probes for bacteria and archaea (EUB338-I-III and ARCH968, respectively). Generally, the bacteria (Figures 5.4a-5.4c) were more abundant than the archaea (Figures 5.4d-5.4f) with a ratio of 9:1. The archaeal population was estimated around 10% after the enrichment; however, to demonstrate the possible occurrence of AOM, anaerobic methanotrophs were specifically targeted among the different microbial community. Therefore, the HRP-labelled oligonucleotide probe that targets ANME-2 (ANME-2-538) was used, which is the ANME type usually observed in other seep sediments (Vigneron et al., 2013) and AOM enrichments in bioreactors (Meulepas et al., 2009; Timmers et al., 2015; Zhang et al., 2011).

In the BTF biomass fed with methane for 213 days, only a few ANME-2 cells not larger than 1 µm were observed under the microscope (Figures 5.4g-5.4i). The ANME-2 cells visualized were surrounded by other microorganisms (DAPI staining in blue, Figure 5.4). ANME-2 are usually associated with DSS (Knittel and Boetius, 2009; Schreiber et al., 2010) and thus, the microorganisms surrounding the ANME-2 cells might be the DSS bacterial partner. The cooperative interaction between the ANME-2 and DSS is still under debate: recent studies have showed interactions by direct electron transfer (McGlynn et al., 2015; Wegener et al., 2016), while another study showed that they can be decoupled by using external electron acceptors (Scheller et al., 2016).

The DSS cells were visualized prior to inoculation and at the end of the BTF operation by using the DSS658 HRP-labelled probe (Figure 5.5). More than 70% of the cells were stained with the DSS probe. Most of the DSS cells had a vibrioid morphology and were 3-5 µm long. The DSS cells found in association with ANME-2 are usually coccoid or rod-shaped (Knittel and Boetius 2009), only few studies have indicated vibrio-shaped DSS associated with ANME-2 (Schreiber et al. 2010). The shape variations of the DSS reveal the genomic difference of the microorganisms (Schreiber et al., 2010). It has to be noted that DSS embrace different phylogenetic and metabolic subgroups of SRB and the DSS subgroup SEEP-SRB1 was identified as the bacterial partner of ANME-2 (Schreiber et al 2010). However, even in the SEEP-SRB1 cluster, microorganisms with different shape and genome can be easily noticed.

The DSS found at the end of the BTF operation were higher in number (~70%, Figures 5.5a-5.5f) when compared to the original sediment (~40%, Figures 5.5g and 5.5l), which clearly suggests that they have been enriched during BTF operation by feeding thiosulfate. Specifically, mainly the vibrio-shaped DSS were enriched in the BTF (Figures 5.5a-5.5f). Before the BTF inoculation, the DSS cells were smaller in size (1 to 3 µm) and more morphologically diverse: coccoid (Figures 5.5h, 5.5i, 5.5j, 5.5l), rod-shaped (Figures 5.5g, 5.5h, 5.5i, 5.5k) and vibrio-shaped (Figures 5.5g and 5.5h). Coccoid cells were mostly found in aggregates with other microorganisms (Figures 5.5h and 5.5l), while the vibrio-shaped were preferentially alone (Figures 5.5g and 5.5h). After the enrichment in the thiosulfate fed BTF only the vibrioid cells were visualized, either in aggregates with other cells (Figures 5.5a and 5.5b), but mainly distant from other microbes (Figures 5.5c-5.5f). Probably, these vibrio-shaped DSS cells are the ones responsible for the disproportionation of thiosulfate to sulfide and sulfate and they probably did not require any partner. Nonetheless, it is unclear if they are also able to reduce sulfate and function as partner for ANME.

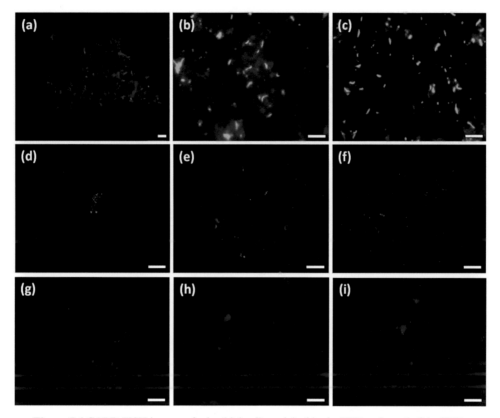

**Figure 5.4** CARD-FISH images of microbial cells enriched in the BTF at the end of the BTF operation (213 days). All panels show confocal laser scanning micrographs: (a-c) DAPI stained cells in blue and EUB338-I-III stained cells in light blue, (d-f) ARCH968 stained cells in orange and (g-i) ANME-2-538 stained cells in red. Scale bars represent 10 µm.

DSS were only once described as disproportionating bacteria in the literature (Milucka et al., 2012). In this study, it was hypothesized that methane was oxidized to bicarbonate and sulfate was reduced to zero-valent sulfur (as an intracellular intermediate) by ANME-2 cells. The resulting sulfur was then released outside the cell as disulfide, which was supposedly disproportionated into sulfide and sulfate by the bacterial partner DSS. Moreover, in the same study vibrio-shaped DSS, similar to our study, were visualized by CARD-FISH after 70 days incubation with colloidal sulfur, showing disproportionation of zero-valent sulfur to sulfide and sulfate, respectively (Milucka et al., 2012).

Jagersma et al. (2012) showed that ANME-1 cells were enriched in incubations with marine Eckernförde Bay sediment in the presence of methane and thiosulfate, but no other ANME types or DSS were enriched. In contrast, ANME-2 cells were detected in this study using Alpha Mound sediment as inoculum, but we cannot exclude the presence of other ANME types (i. e. ANME-1 or ANME-3) since only the probe targeting ANME-2 was used in this study.

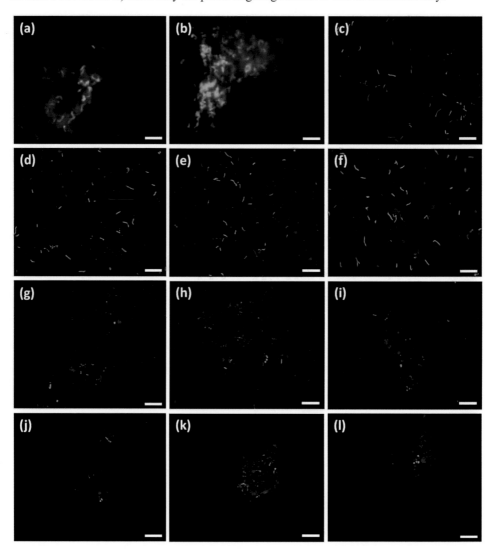

**Figure 5.5** CARD-FISH images of microbial cells stained with DAPI in blue and stained with DSS658 probe in green. All panels show confocal laser scanning micrographs of cells (a-f) enriched in the BTF after 213 days operation and (g-l) before inoculation in the BTF. Scale bars represent 10 μm.

To our knowledge, no DSS have been described to disproportionate thiosulfate as it occurred in the BTF operated in this study. Recently, high rates of thiosulfate disproportionation have been reported in a study with marine sediment and methane as the sole carbon source (Suarez-Zuluaga et al., 2014). In that study, a high number of *Desulfocapsa* was observed, a bacterium specialized in the disproportionation of sulfur compounds (Finster, 2008). Many SRB can metabolize inorganic sulfur compounds by disproportionation: *Desulfocapsa sulfoexigens* is specialized, and also other SRB such as *Desulfovibrio desulfodismutans* and *Desulfocapsa thiozymogenes* can disproportionate thiosulfate (Finster, 2008). These microorganisms belong to the families of, respectively, the *Desulvovibrionaceae* and *Desulfobulbaceae,* which were found to be absent in the BTF sludge, and not to the *Desulfobacteraceae* to which DSS belong.

The BTF was suitable for the enrichment of slow growing microorganisms: the reactor did not clog and the biomass was easily kept active during the long term reactor operation. The sulfate reducing bacteria community was enriched and maintained with the use of thiosulfate even if only methane was supplied as electron donor. For the application of this process for the desulfurization of wastewaters, such as mining and metallurgical waste streams, it is necessary to quantify and establish the mode of sulfide reoxidation and other concomitant side reactions. Moreover, it is necessary to quantify the removal of the greenhouse gas methane through anaerobic methane oxidation.

Further research is needed to identify the specific group of DSS able to perform the disproportionation of thiosulfate and to further explore the possibility of this type of bacteria to function as partner for ANME-2. Moreover, the occurrence of net AOM concomitant to thiosulfate disproportionation and upon the formation of sulfate by the latter process could be studied in more detail by using isotopic labeled methane (Timmers et al., 2015) or by FISH coupled to microautoradiography (Lee et al., 1999). For possible future applications, the enriched DSS biomass could be used as inoculum for AOM-SR in bioreactors.

## 5.4    Conclusions

Long-term operation of a BTF fed with thiosulfate and methane showed that thiosulfate disproportionation to sulfate and sulfide prevailed over its reduction with methane as sole electron donor. In the absence of thiosulfate, the formed sulfate was readily reduced coupled to the oxidation of methane. The use of polyurethane foam as a packing material for the BTF and the addition of thiosulfate decreased the start-up time required for sulfate reduction, with the highest sulfide production rate being 0.5 mmol $l^{-1}$ day$^{-1}$. ANME-2 cells were hardly present in the enrichment, while DSS were highly enriched and probably responsible for thiosulfate disproportionation.

## 5.5    References

Acree, T.E., Sonoff, E.P., Splittstoesser, D.F., 1971. Determination of hydrogen sulfide in fermentation broths containing SO$_2$. *Appl. Microbiol.*, **22**(1), 110-112.

Aoki, M., Ehara, M., Saito, Y., Yoshioka, H., Miyazaki, M., Saito, Y., Miyashita, A., Kawakami, S., Yamaguchi, T., Ohashi, A., Nunoura, T., Takai, K., Imachi, H. 2014. A long-term cultivation of an anaerobic methane-oxidizing microbial community from deep-sea methane-seep sediment using a continuous-flow bioreactor. *PLoS ONE*, **9**(8), Pe105356.

Cypionka, H., Smock, A.M., Böttcher, M.E., 1998. A combined pathway of sulfur compound disproportionation in *Desulfovibrio desulfuricans*. *FEMS Microbiol. Ecol.*, **166**(2), 181-186.

Daims, H., Brühl, A., Amann, R., Schleifer, K.-H., Wagner, M. 1999. The domain-specific probe EUB338 is insufficient for the detection of all Bacteria: development and evaluation of a more comprehensive probe set. *Syst. Appl. Microbiol.*, **22**(3), 434-444.

Egger, M., Rasigraf, O., Sapart, C.l.J., Jilbert, T., Jetten, M.S., Röckmann, T., van der Veen, C., Banda, N., Kartal, B., Ettwig, K.F., Slomp, C.P. 2015. Iron-mediated anaerobic oxidation of methane in brackish coastal sediments. *Environ. Sci. Technol.*, **49**(1), 277-283.

Ettwig, K.F., Butler, M.K., Le Paslier, D., Pelletier, E., Mangenot, S., Kuypers, M.M.M., Schreiber, F., Dutilh, B.E., Zedelius, J., de Beer, D., Gloerich, J., Wessels, H.J.C.T., van Alen, T., Luesken, F., Wu, M.L., van de Pas-Schoonen, K.T., Op den Camp, H.J.M., Janssen-Megens, E.M., Francoijs, K.-J., Stunnenberg, H., Weissenbach, J., Jetten, M.S.M., Strous, M. 2010. Nitrite-driven anaerobic methane oxidation by oxygenic bacteria. *Nature*, **464**(7288), 543-548.

Estrada, J.M., Lebrero, R., Quijano, G., Pérez, R., Figueroa-González, I., García-Encina, P.A., Muñoz, R., 2014. Methane abatement in a gas-recycling biotrickling filter: evaluating innovative operational strategies to overcome mass transfer limitations. *Chem. Eng. J.*, **253**(53), 385-393.

Finster, K. 2008. Microbiological disproportionation of inorganic sulfur compounds. *J. Sulfur Chem.*, **29**(3-4), 281-292.

Finster, K., Liesack, W., Thamdrup, B. 1998. Elemental sulfur and thiosulfate disproportionation by *Desulfocapsa sulfoexigens* sp . nov., a new anaerobic bacterium isolated from marine surface sediment. *Appl. Environ. Microb.*, **64**(1), 119-125.

Flynn, T.M., O'Loughlin, E.J., Mishra, B., DiChristina, T.J., Kemner, K.M., 2014. Sulfur-mediated electron shuttling during bacterial iron reduction. *Science*. **344**(6187), 1039-42.

Forster, A., Schouten, S., Baas, M., Sinninghe Damsté, J.S., 2007. Mid-Cretaceous (Albanian-Santonian) sea surface temperature record of the tropical. *Atlantic Ocean. Geology*. **35**(10), 919-919.

Gonzalez-Gil, G., Meulepas, R.J.W., Lens, P.N.L. 2011. Biotechnological aspects of the use of methane as electron donor for sulfate reduction. in: Murray, M.-Y. (Ed.), *Comprehensive Biotechnology*, Vol. 6 (2$^{nd}$ edition), Elsevier B.V. Amsterdam, the Netherlands, pp. 419-434.

.Liamleam, W., Annachhatre, A.P. 2007. Electron donors for biological sulfate reduction. *Biotechnol. Adv.*, **25**(5), 452-463.

Holler, T., Wegener, G., Niemann, H., Deusner, C., Ferdelman, T.G., Boetius, A., Brunner, B., Widdel, F. 2011. Carbon and sulfur back flux during anaerobic microbial oxidation of methane and coupled sulfate reduction. *Proc. Natl. Acad. Sci. USA*, **108**(52), E1484-E1490.

Jagersma, C.G., Meulepas, R.J.W., Timmers, P.H.A., Szperl, A., Lens, P.N.L., Stams, A.J.M., 2012. Enrichment of ANME-1 from Eckernförde Bay sediment on thiosulfate, methane and short-chain fatty acids. *J. Biotechnol.*, **157**(4), 482-9.

Kamyshny, A., Gun, J., Rizkov, D., Voitsekovski, T., Lev, O., 2007. Equilibrium distribution of polysulfide ions in aqueous solutions at different temperatures by rapid single phase derivatization. *Environ. Sci. Technol.*, **41**(7), 2395-2400.

Kirschke, S., Bousquet, P., Ciais, P., Saunois, M., Canadell, J.G., Dlugokencky, E.J., Bergamaschi, P., Bergmann, D., Blake, D.R., Bruhwiler, L., Cameron-Smith, P., Castaldi, S., Chevallier, F., Feng, L., Fraser, A., Heimann, M., Hodson, E.L., Houweling, S., Josse, B., Fraser, P.J., Krummel, P.B., Lamarque, J.-F., Langenfelds, R.L., Le Quere, C., Naik, V., O'Doherty, S., Palmer, P.I., Pison, I., Plummer, D., Poulter, B., Prinn, R.G., Rigby, M., Ringeval, B., Santini, M., Schmidt, M., Shindell, D.T., Simpson, I.J., Spahni, R., Steele, L.P., Strode, S.A., Sudo, K., Szopa, S., van der Werf, G.R., Voulgarakis, A., van Weele, M., Weiss, R.F., Williams, J.E., Zeng, G. 2013. Three decades of global methane sources and sinks. *Nat. Geosci.*, **6**(10), 813-823.

Knittel, K., Boetius, A. 2009. Anaerobic oxidation of methane: progress with an unknown process. *Annu. Rev. Microbiol.*, **63**(1), 311-334.

Lee, N., Nielsen, P.H., Andreasen, K.H., Juretschko, S., Nielsen, J.L., Schleifer, K.-H., Wagner, M. 1999. Combination of fluorescent *in situ* hybridization and microautoradiography-a new tool for structure-function analyses in microbial ecology. *Appl. Environ. Microbiol.*, **65**(3), 1289-1297.

Liamleam, W., Annachhatre, A.P. 2007. Electron donors for biological sulfate reduction. *Biotechnol. Adv.*, **25**(5), 452-463.

Maignien, L., Depreiter, D., Foubert, a., Reveillaud, J., Mol, L., Boeckx, P., Blamart, D., Henriet, J.P., Boon, N., 2010. Anaerobic oxidation of methane in a cold-water coral carbonate mound from the Gulf of Cadiz. *Int. J. Earth Sci.*, **100**(6), 1413-1422.

Manz, W., Eisenbrecher, M., Neu, T.R., Szewzyk, U., 1998. Abundance and spatial organization of gram-negative sulfate-reducing bacteria in activated sludge investigated by *in situ* probing with specific 16S rRNA targeted oligonucleotides. *FEMS Microbiol. Ecol.*, **25**(1), 43-61.

McGlynn, S.E., Chadwick, G.L., Kempes, C.P., Orphan, V.J. 2015. Single cell activity reveals direct electron transfer in methanotrophic consortia. *Nature*, **526**(7574), 531-535.

Meulepas, R.J.W., Jagersma, C.G., Gieteling, J., Buisman, C.J.N., Stams, A.J.M., Lens, P.N.L. 2009. Enrichment of anaerobic methanotrophs in sulfate-reducing membrane bioreactors. *Biotechnol. Bioeng.*, **104**(3), 458-470.

Meulepas, R.J.W., Stams, A.J.M., Lens, P.N.L. 2010. Biotechnological aspects of sulfate reduction with methane as electron donor. *Rev. Environ. Sci. Biotechnol.*, **9**(1), 59-78.

Milucka, J., Ferdelman, T.G., Polerecky, L., Franzke, D., Wegener, G., Schmid, M., Lieberwirth, I., Wagner, M., Widdel, F., Kuypers, M.M.M. 2012. Zero-valent sulphur is a key intermediate in marine methane oxidation. *Nature*, **491**(7425), 541-546.

Niemann, H., Losekann, T., de Beer, D., Elvert, M., Nadalig, T., Knittel, K., Amann, R., Sauter, E.J., Schluter, M., Klages, M., Foucher, J.P., Boetius, A. 2006. Novel microbial communities of the Haakon Mosby mud volcano and their role as a methane sink. *Nature*, **443**(7113), 854-858.

Pernthaler Annelie , P.J., Amann Rudolf. 2002. Fluorescence *in situ* hybridization and catalyzed reporter deposition for the identification of marine bacteria. *Appl. Environ. Microbiol.*, **68**(6), 3094-3101.

Pernthaler, A., Pernthaler, J., Amann, R., 2004. Sensitive multi-color fluorescence *in situ* hybridization for the identification of environmental microorganisms, in: Kowalchuk, G., de Bruijn, F., Head, I., Akkermans, A., van Elsas, J.D. (Eds.), *Molecular Microbial Ecology Manual.* Springer, Dordrecht, pp. 711-726.

Poser, A., Lohmayer, R., Vogt, C., Knoeller, K., Planer-Friedrich, B., Sorokin, D., Richnow, H.H., Finster, K. 2013. Disproportionation of elemental sulfur by haloalkiphilic bacteria from soda lakes. *Extremophiles* **17**(6), 1003-1012.

Scheller, S., Yu, H., Chadwick, G.L., McGlynn, S.E., Orphan, V.J. 2016. Artificial electron acceptors decouple archaeal methane oxidation from sulfate reduction. *Science*, **351**(6274), 703-707.

Schreiber, L., Holler, T., Knittel, K., Meyerdierks, A., Amann, R. 2010. Identification of the dominant sulfate-reducing bacterial partner of anaerobic methanotrophs of the ANME-2 clade. *Environ. Microbiol.*, **12**(8), 2327-2340.

Sekar, R., Pernthaler, A., Pernthaler, J., Posch, T., Amann, R., Warnecke, F., 2003. An improved protocol for quantification of freshwater *Actinobacteria* by fluorescence *in situ* hybridization. *Appl. Environ. Microbiol.*, **69**(5), 2928-2935.

Stahl, D.A., Amann, R.I. 1991. Development and application of nucleic acid probes. In: Stackebrandt, E., Goodfellow, M. (Eds.), *Nucleic acid techniques in bacterial systematics.* John Wiley & Sons Ltd, Chichester, UK, pp. 205-248.

Suarez-Zuluaga, D.A., Timmers, P.H.A., Plugge, C.M., Stams, A.J.M., Buisman, C.J.N., Weijma, J. 2015. Thiosulphate conversion in a methane and acetate fed membrane bioreactor. *Environ. Sci. Pollut. Res.*, **23**(3), 2467-2478.

Suarez-Zuluaga, D.A., Weijma, J., Timmers, P.H.A., Buisman, C.J.N. 2014. High rates of anaerobic oxidation of methane, ethane and propane coupled to thiosulphate reduction. *Environ. Sci. Pollut. Res.*, **22**(5), 3697-3704.

Templer, S.P., Wehrmann, L.M., Zhang, Y., Vasconcelos, C., McKenzie, J.A. 2011. Microbial community composition and biogeochemical processes in cold-water coral carbonate mounds in the Gulf of Cadiz, on the Moroccan margin. *Mar. Geol.*, **282**(1-2), 138-148.

Thamdrup, B., Fossing, H., Jørgensen, B.B., 1994. Manganese, iron, and sulfur cycling in a coastal marine sediment, Aarhus Bay, Denmark. *Geochim. Cosmochim. Ac.*, **58**(23), 5115-5129.

Timmers, P.H., Gieteling, J., Widjaja-Greefkes, H.A., Plugge, C.M., Stams, A.J., Lens, P.N.L., Meulepas, R.J. 2015. Growth of anaerobic methane-oxidizing archaea and sulfate-reducing bacteria in a high-pressure membrane capsule bioreactor. *Appl. Environ. Microbiol.*, **81**(4), 1286-1296.

Treude, T., Orphan, V.J., Knittel, K., Gieseke, A., House, C.H., Boetius, A. 2007. Consumption of methane and $CO_2$ by methanotrophic microbial mats from gas seeps of the anoxic Black Sea. *Appl. Environ. Microbiol.*, **73**(7), 2271-2283.

Vigneron, A., Cruaud, P., Pignet, P., Caprais, J.-C., Cambon-Bonavita, M.-A., Godfroy, A., Toffin, L. 2013. Archaeal and anaerobic methane oxidizer communities in the Sonora Margin cold seeps, Guaymas Basin (Gulf of California). *ISME J.*, **7**(8), 1595-1608.

Villa-Gomez, D., Ababneh, H., Papirio, S., Rousseau, D.P.L., Lens, P.N.L. 2011. Effect of sulfide concentration on the location of the metal precipitates in inversed fluidized bed reactors. *J. Hazard. Mater.* **192**(1), 200-207.

Wan, M., Shchukarev, A., Lohmayer, R., Planer-Friedrich, B., Peiffer, S. 2014. Occurrence of surface polysulfides during the interaction between ferric (hydr)oxides and aqueous sulfide. *Environ. Sci. Technol.*, **48**(9), 5076-5084.

Wegener, G., Krukenberg, V., Ruff, S.E., Kellermann, M.Y., Knittel, K. 2016. Metabolic capabilities of microorganisms involved in and associated with the anaerobic oxidation of methane. *Front. Microbiol.*, **7**(46), 1-16.

Wehrmann, L.M., Risgaard-Petersen, N., Schrum, H.N., Walsh, E.A., Huh, Y., Ikehara, M., Pierre, C., D'Hondt, S., Ferdelman, T.G., Ravelo, A.C., Takahashi, K., Zarikian, C., 2011. Coupled organic and inorganic carbon cycling in the deep subseafloor sediment of the north-eastern bering sea slope. *Chem. Geol.*, **284**(3-4), 251-261.

Weijma, J., Veeken, A., Dijkman, H., Huisman, J., Lens, P.N.L. 2006. Heavy metal removal with biogenic sulphide: advancing to full-scale. in: Cervantes, F., Pavlostathis, S., van Haandel, A. (Eds.), *Advanced biological treatment processes for industrial wastewaters, principles and applications*, IWA publishing. London, pp. 321-333.

Wendeberg, A., 2010. Fluorescence *in situ* hybridization for the identification of environmental microbes. *Cold Spring Harb. Protoc.,* **5**(1), 1-9.

Widdel, F., Bak, F. 1992. Gram negative mesophilic sulfate reducing bacteria. in: Balows, A., Truper, H., Dworkin, M., Harder, W., Schleifer, K. H. (Eds.), *The prokaryotes: a handbook on the biology of bacteria: ecophysiology, isolation, identification, applications.*, Vol. 2, Springer New York, USA, pp. 3352-3378.

Zhang, Y., Henriet, J.-P., Bursens, J., Boon, N. 2010. Stimulation of *in vitro* anaerobic oxidation of methane rate in a continuous high-pressure bioreactor. *Biores. Technol.,* **101**(9), 3132-3138.

Zhang, Y., Maignien, L., Zhao, X., Wang, F., Boon, N. 2011. Enrichment of a microbial community performing anaerobic oxidation of methane in a continuous high-pressure bioreactor. *BMC Microbiol.,* **11**(137), 1-8.

# CHAPTER 6

# Enrichment of Anaerobic Methanotrophs in Biotrickling Filters using Different Sulfur Compounds as Electron Acceptors

**Abstract**

A biotrickling filter (BTF) operating at ambient pressure and temperature was used to enrich microorganisms from a deep sea anaerobic methane oxidizing sediment (Alpha Mound, Gulf of Cadiz). Different sulfur compounds namely, sulfate, elemental sulfur and thiosulfate were used as electron acceptors to understand their effects on the anaerobic oxidation of methane (AOM), sulfate reduction rates and the microbial community distribution. The highest AOM and sulfate reduction rates were obtained in the BTF fed with thiosulfate as the electron acceptor (~0.4 mmol l$^{-1}$ day$^{-1}$). The use of thiosulfate triggered the enrichment of sulfate reducing bacteria (SRB) in the BTF, while the highest number of anaerobic methanotrophs (ANME) was visualized in the sulfate fed BTF (ANME-2 43% of the total visualized archaea), where sulfate was reduced at a maximum rate of 0.3 mmol l$^{-1}$ day$^{-1}$. This study shows that ANME and SRB obtained from deep sea conditions (528 m below sea level) can be enriched in a BTF at ambient pressure and temperature with a relatively short start-up time (42 days).

## 6.1 Introduction

Anaerobic oxidation of methane (AOM) coupled to sulfate reduction (SR) is a biological process occurring in anoxic environments, especially in marine sediments (Reeburgh, 2007; Knittel and Boetius, 2009; Scheller et al., 2016). AOM contributes to the removal of methane, thereby controlling its emission to the atmosphere (Hinrichs and Boetius, 2002; Raghoebarsing et al., 2006; Reeburgh, 2007). Methane is a well-known greenhouse gas and its presence in the atmosphere at high concentrations has large implications for future climate change (Forster et al., 2007). Many terrestrial and aquatic surfaces are possible methane sources, thus, it is important to understand the processes and mechanisms involved in its consumption (Kirschke et al., 2013).

AOM coupled to SR (AOM-SR) is a process mediated by anaerobic methanotrophs (ANME) and sulfate reducing bacteria (SRB). ANME are grouped into three distinct clades, i.e. ANME-1, ANME-2 and ANME-3, respectively (Hinrichs et al., 1999; Orphan et al., 2001; Knittel and Boetius, 2009; Bhattarai et al. 2017a). The common SRB associated with ANME are *Desulfosarcina/Desulfococcus* (DSS) and *Desulfobubaceae* (DBB) (Schreiber et al., 2010). Understanding the mechanism of this process has always been a challenge due to the difficulty in enriching the ANME under laboratory conditions. These archaea have not yet been isolated in pure culture and they are extremely slow growing organisms, having a doubling time of ~2 to 7 months (Girguis et al., 2005; Nauhaus et al., 2007; Krüger et al., 2008; Deusner et al., 2009; Meulepas et al., 2009a; Zhang et al., 2011; Wegener et al. 2016). In addition, ANME require strict anaerobic conditions and high methane availability, which is rather difficult to achieve at laboratory conditions due to the low solubility of methane in water at standard atmospheric pressure and temperature (1.3 mM in seawater at 20°C). Theoretically, elevated methane partial pressure favors AOM-SR, as more methane will be dissolved. ANME are usually found in deep sea sediments, where the pressure and temperature range significantly from these observed at ambient conditions, e.g. Gulf of Cadiz sediment is subjected to pressures higher than 10 MPa and temperatures lower than 10°C (Niemann et al., 2006). Such environmental conditions are difficult to simulate in the laboratory.

The enrichment of ANME can be enhanced by the use of different types of bioreactor configurations such as a high pressure reactor (Deusner et al., 2009; Zhang et al., 2011), a membrane reactor (Meulepas et al., 2009a; Timmers et al., 2015a) or a biotrickling filter (Cassarini et al. 2017). However, the SR rates reported so far (~0.6 mmol l$^{-1}$day$^{-1}$) with methane as the electron donor are more than 100 times lower than the rates achieved with other electron donors, such as hydrogen or ethanol (Suarez-Zuluaga et al., 2014; Bhattarai et al., 2017b). The highest specific AOM rate (370 μmol g dry weight$^{-1}$ day$^{-1}$) has been obtained with sediment from the Black Sea microbial mat as inoculum in a high pressure bioreactor incubated at a methane partial pressure of 6 MPa and at 20°C (Deusner et al., 2009). At ambient pressure, the highest volumetric SR rate (0.6 mmol l$^{-1}$ day$^{-1}$) was reported by Meulepas et al. (2009a) in a 2 l membrane bioreactor operated for 884 days. However, that bioreactor required a long start-up period of ~400 days.

In a recent study, Cassarini et al. (2017) operated a biotrickling filter (BTF) for 213 days with the sediment collected from the Alpha Mound (Gulf of Cadiz, Spain) as inoculum and showed AOM coupled to thiosulfate reduction. The BTF was operated at ambient conditions, using porous polyurethane foam as the packing material. The DSS population was enriched in the BTF, while the sulfide production rates increased (from 0.01 to 0.5 mmol l$^{-1}$ day$^{-1}$) and the SR (0.4 mmol l$^{-1}$ day$^{-1}$) was immediately activated after complete consumption of thiosulfate with methane as the sole electron donor. Besides, other advantages of using a BTF compared to high pressure bioreactors are the enhanced gas-liquid mass transfer in the filter bed, better gas-liquid mixing characteristics, flexibility in reactor operation (up-flow or down-flow modes), ease of reactor maintenance, and low operational and maintenance costs. The polyurethane foam cubes, used as packing material of the BTF, are highly porous and the methane is partly retained in the pores increasing the gas-liquid mass transfer (Aoki et al., 2014; Estrada et al., 2014a), while the biomass attaches onto the packing material facilitating its growth. However, the ANME were scarcely present and the AOM rates could not be determined since the methane consumed and carbon dioxide produced solely by AOM could not be determined by the methods previously used (Cassarini et al., 2017).

AOM coupled to thiosulfate reduction is thermodynamically more favorable than AOM coupled to SR (Table 6.1, Eqs. 6.1 and 6.2), but also elemental sulfur could be used as electron acceptor for methane oxidation since it can presumably be used directly by some ANME clades (Milucka et al., 2012). The aim of this study was, therefore, to investigate the effect of different sulfur compounds used as substrates for ANME and SRB at ambient pressure and temperature in a BTF. In this study, sulfate and elemental sulfur were used as electron acceptors in two identical BTF, similar to the reactor described in Cassarini et al. (2017), which operated with thiosulfate as electron acceptor. The biomass enriched after 230 and 147 days of operation in the two BTF operating in parallel as well as the biomass from the thiosulfate fed BTF after 213 days (Cassarini et al., 2017) was used in batch activity assays using $^{13}$C-labelled methane ($^{13}$CH$_4$) to investigate AOM and determine the AOM rate in the presence of different sulfur compounds as electron acceptors.

**Table 6.1** Sulfate, thiosulfate and elemental sulfur reduction and thiosulfate and elemental sulfur disproportionation reactions.

| Eq. | Reactions | $\Delta_r G°$ | $\Delta_r G'$ |
|---|---|---|---|
| 6.1 | $CH_4 + SO_4^{2-} \rightarrow HCO_3^- + HS^- + H_2O$ | -16.6 kJ mol$^{-1}$ CH$_4$ | -25.7 kJ mol$^{-1}$ CH$_4$ |
| 6.2 | $CH_4 + S_2O_3^{2-} \rightarrow HCO_3^- + 2HS^- + H^+$ | -38.5 kJ mol$^{-1}$ CH$_4$ | -64.5 kJ mol$^{-1}$ CH$_4$ |
| 6.3 | $CH_4 + 4S^0 + 3H_2O \rightarrow HCO_3^- + 4HS^- + 5H^+$ | +24.3 kJ mol$^{-1}$ S$^0$ | Not determined |
| 6.4 | $S_2O_3^{2-} + H_2O \rightarrow HS^- + SO_4^{2-} + H^+$ | -21.9 kJ mol$^{-1}$ S$_2$O$_3$$^{2-}$ | -41.3 kJ mol$^{-1}$ S$_2$O$_3$$^{2-}$ |
| 6.5 | $4S^0 + 4H_2O \rightarrow 3HS^- + SO_4^{2-} + 5H^+$ | +40.9 kJ mol$^{-1}$ S$^0$ | Not determined |

**Note:**

Gibbs free energy of reactions at standard conditions ($\Delta_r G°$) were obtained from Thauer et al. 1977 and under the following operational conditions of the BTF ($\Delta_r G'$): pH 7.0, CH$_4$ 1.27 mM, HCO$_3^-$ 30 mM, SO$_4^{2-}$ 10 mM, S$_2$O$_3^{2-}$ 10 mM and HS$^-$ 0.01 mM.

## 6.2 Material and methods

### 6.2.1 Source of biomass and composition of artificial seawater medium

Sediment samples were obtained from the Alpha Mound (35°17.48'N; 6°47.05'W, water depth ca. 525 m), Gulf of Cadiz (Spain), during the R/V Marion Dufresne Cruise MD 169 MiCROSYSTEMS to the Gulf of Cadiz in July 2008. The characteristics of the sediment have been described in Cassarini et al. (2017), while the preparation procedure and the composition of the artificial seawater medium have been described in Bhattarai et al. (2017a). The vitamins and trace element mixtures were prepared according to Widdel and Bak (1992). 0.5 g l$^{-1}$ resazurin solution was added as the redox indicator and 0.01 mM of sodium sulfide was added as the reducing agent to the seawater medium. The pH of the seawater medium was adjusted to 7.0 with sterile 1 M Na$_2$CO$_3$ or 1 M H$_2$SO$_4$ solution. The medium was maintained under anoxic conditions with the help of nitrogen purging until it was recirculated to the three BTF.

All the chemicals were purchased as lab grade from Fisher Scientific (Sheepsbouwersweg, the Netherlands). Na$_2$SO$_4$ and Na$_2$S$_2$O$_3$ were used in their anhydrous form, while elemental sulfur was used as precipitated sulfur powder (Fisher Scientific, Sheepsbouwersweg, the Netherlands). In the S$^0$ BTF, 50 g of elemental sulfur was added together with the sediment. In the $SO_4^{2-}$ BTF, $SO_4^{2-}$ was added to the artificial seawater medium as $Na_2SO_4$ (1.4 g per liter of demineralised water), while $Na_2S_2O_3$ (1.6 g per liter of demineralised water) was added in the $S_2O_3^{2-}$ BTF, as described previously (Cassarini et al., 2017).

## 6.2.2 BTF setup and operation

The three BTF were operated in parallel to investigate AOM with different electron acceptors, namely sulfate, elemental sulfur or thiosulfate. They were maintained in the dark and at room temperature ($\sim 20 \pm 2$ °C). The three BTFs were identical, constructed using acrylic cylinders and sealed air-tight to prevent leakage or air intrusion during its operation. The filter bed volume of each BTF was 0.4 l and polyurethane foam cubes (BVB Sublime, the Netherlands) of 1 cm$^3$ (void fraction of 0.98 and density of 28 kg m$^{-3}$) were used as the packing material.

The three BTF were operated in sequential fed-batch mode for the artificial seawater medium, while the gas-phase methane (99.5% methane, Linde gas, Schiedam, the Netherlands) was stored in air tight Tedlar bags (Sigma-Aldrich, USA) and supplied to the BTF. During BTF start-up, 20 ml of Alpha Mound Sediment (0.03 $\pm$ 0.01 g volatile suspended solids) was inoculated to each BTF.

The BTF with sulfate as electron acceptor ($SO_4^{2-}$ BTF) was operated for 230 days and the seawater medium containing 10 mM sulfate was replaced periodically on days 49, 142, 178, 197 and 217, respectively. The operation of the $SO_4^{2-}$ BTF was arbitrarily divided in different phases, each one ending before medium replacement: days 0-41 (I), days 42-139 (II), days 140-174 (III), days 175-190 (IV), days 191-217 (V), days 218-230 (VI) (Figure 6.1). For each phase, the volumetric SR and the total dissolved sulfide production rates were determined. Biomass samples for microbial visualization by catalyzed reporter deposition-fluorescence *in situ* hybridization (CARD-FISH) were obtained before inoculation, and on days 84, 155 and at the end of the BTF operation (day 230).

The BTF for the investigation of elemental sulfur as electron acceptor ($S^0$ BTF) was operated for 147 days and the pH was adjusted to 7.0 with sterile 1 M $Na_2CO_3$ on days 80, 105, 126 and 137, whenever the pH dropped below 5.0. The operation of the $S^0$ BTF was arbitrarily divided in two phases (Phase I ending before any pH adjustment): days 0-74 (I), days 75-147 (II) (Figure 6.1). For each phase, the volumetric SR and the total dissolved sulfide production rates were determined.

The BTF for the investigation of thiosulfate as electron acceptor ($S_2O_3^{2-}$ BTF) has been described in Cassarini et al. (2017) and after 213 days of operation, the biomass was transferred in 118 ml serum bottles for performing activity assays and determining the AOM rates. For each BTF, both gas (inlet and outlet) and liquid effluent samples were collected twice a week from the sampling ports (Cassarini et al., 2017). pH, sulfate, sulfide and thiosulfate concentrations were measured in samples collected from the liquid medium, while methane and carbon dioxide were analyzed from the gas samples collected at the inlet and outlet of the BTF (Cassarini et al., 2017).

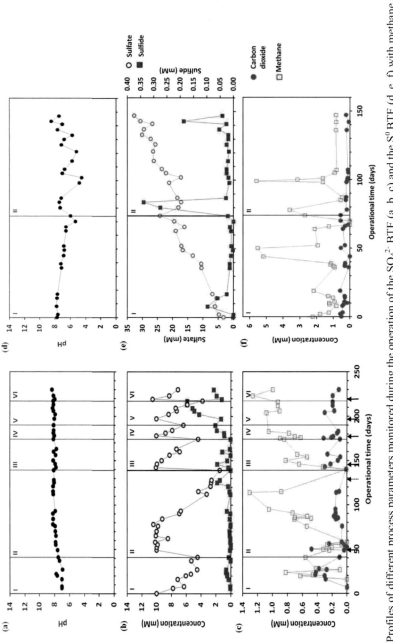

**Figure 6.1** Profiles of different process parameters monitored during the operation of the $SO_4^{2-}$ BTF (a, b, c) and the $S^0$ BTF (d, e, f) with methane as the electron donor and sulfate as the electron acceptor: pH (a, d), sulfate and sulfide (b, e), and methane and carbon dioxide concentration measured as [methane inlet- methane outlet] and [carbon dioxide outlet - carbon dioxide inlet] (c, f). The vertical lines represent the different phases of the BTF operation. The black arrows indicate the days at which the mineral medium was replaced, while the dashed arrows indicate the days at which yellow coloration of the mineral medium was observed.

## 6.2.3   Activity assays

The occurrence of AOM and the estimation of the methane oxidation rates were investigated from the $^{13}C$-labeled carbon dioxide produced during batch tests with $^{13}CH_4$. The polyurethane foam cubes containing the enriched inoculum were collected from each BTF at the end of their operation. The tests were performed in triplicate to evaluate the standard deviation. Control batch were prepared under nitrogen (without methane), in the absence of the electron acceptor and without the biomass (fresh polyurethane foam cubes were added). Ten polyurethane foam cubes containing the BTF biomass were added to each batch incubation done in previously weighted 118 ml serum bottles, which were then immediately closed with gas-tight butyl rubber stoppers and capped. To ensure strict anaerobic conditions in the bottles and avoid any oxygen intrusion, the gas phase was replaced several times with nitrogen gas and made vacuum thereafter. 52 ml of anaerobic artificial seawater was added to the bottles and the headspace was flushed with methane for 8 min. An estimated equivalent amount of 5% of headspace was taken out and once again filled with the same amount of $^{13}CH_4$.

The bottles were placed on an orbital shaker (Cole-Parmer, Germany) at 100 rpm in the dark at the operation temperature of the BTF ($22 \pm 3°C$) for 35 days and sampling was performed once a week for both gas and liquid phase analysis. Biomass samples for microbial visualization by CARD-FISH were withdrawn at the end of the activity assays, i.e. on day 35.

## 6.2.4   Chemical analysis

The analysis of pH, sulfate, thiosulfate and dissolved sulfide concentrations were performed in duplicate. The pH, sulfate, thiosulfate, methane and carbon dioxide concentrations were analyzed according to the procedure described in Cassarini et al. (2017). The total dissolved sulfide concentrations ($H_2S$, $HS^-$ and $S^{2-}$) in the three BTF were analyzed spectrophotometrically using the methylene blue method (Acree et al., 1971).

The volatile suspended solids were estimated before inoculation on the basis of the difference between the dry weight total suspended solids and the ash weight of the sediment according to the procedure outlined in Standard Methods (APHA 1995).

The stable carbon isotope composition of methane and carbon dioxide was determined using a gas chromatography - isotope ratio mass spectrometer (GC-IRMS, Agilent 7890A) and the carbon isotopic fraction ($^{13}C/^{12}C$) was estimated as described previously (Dorer et al., 2016). Measurements of the stable isotope composition of $CH_4$ and $CO_2$ were performed in triplicate and the standard deviation was observed to be less than 0.5 δ-units. For quality assurance, standard gas mixtures of methane and carbon dioxide were measured periodically during the entire isotope analysis.

## 6.2.5   Calculations

The volumetric SR and the total dissolved sulfide production rates from the BTF and the activity assays were calculated according to Eqs. 6.6 and 6.7 (Meulepas et al., 2009a):

$$\text{Volumetric SR rate (mmol } l^{-1}d^{-1}) = \frac{[SO_4^{2-}{}_{(t)}]-[SO_4^{2-}{}_{(t+\Delta t)}]}{\Delta t} \qquad \text{Eq. 6.6}$$

$$\text{Volumetric sulfide production rate (mmol } l^{-1} d^{-1}) = \frac{S_{(t+\Delta t)}^{2-}-S_t^{2-}}{\Delta t} \qquad \text{Eq. 6.7}$$

where, $SO_4^{2-}{}_{(t)}$ is the concentration (mmol $l^{-1}$) of sulfate at time (t) during the batch incubation, $SO_4^{2-}{}_{(t+\Delta t)}$ is the concentration (mmol $l^{-1}$) of sulfate at time (t+$\Delta$t), and $S_{(t)}^{2-}$ and $S_{(t+\Delta t)}^{2-}$ are the total dissolved sulfide concentrations at time (t) and at time (t+$\Delta$t), respectively.

The methane oxidation rate was estimated on the basis of the total dissolved inorganic carbon (DIC) produced during the activity assays as described by Dorer et al. (2016). The amount of $^{13}C$-DIC formed at time (t) was indicated as $^{13}C$-DIC(t) and calculated from the measured $\delta^{13}C$ of $CO_2$ (Eq. 6.8).

$$^{13}C\text{-DIC(t) in mmol } l^{-1} = \frac{[DIC_{(start)}]}{1+ R_{VPDB} \times (1+\delta_{(start)})} \times \left(R_{VPDB} \times \left(\delta^{13}C_t - \delta^{13}C_{(start)}\right)\right) \qquad \text{Eq. 6.8}$$

where, $\delta^{13}C_t$ is the isotopic carbon composition relative to the international reference Vienna PeeDeeBelmnite (VPDB) at time (t) of $CO_2$: $\delta^{13}C = \frac{R-R_{VPDB}}{R_{VPDB}}$, where R= $^{13}C/^{12}C$ and $R_{VDPB}$= 0.0112372 (Craig, 1957). $\delta^{13}C_{(start)}$ is the initial isotopic carbon composition of $CO_2$ and $[DIC_{(start)}]$ is the initial concentration of DIC in the incubations i.e. 30 mM. The total amount of DIC formed at time (t) was indicated as DIC(t) and it was calculated as $^{13}C$-DIC(t)/$F_{CH_4(start)}$. $F_{CH_4(start)}$ is the fractional abundance of $^{13}CH_4$ at the start of the incubation defined from $\delta^{13}C$ of $CH_4$ at the beginning ($\delta^{13}CH_{4(start)}$, Eq. 6.9):

$$F_{CH_4(start)} = \frac{R_{VPDB} \times (\delta^{13}CH_{4\ (start)}+1)}{1+R_{VPDB} \times (\delta^{13}CH_{4\ (start)}+1)} \qquad \text{Eq. 6.9}$$

The volumetric AOM rate ($\mu$mol $l^{-1}$ day$^{-1}$) was obtained from the $\Delta$DIC(t)/$\Delta$t values observed during the batch activity tests in which the increase was linear and at least four successive data points were used for its calculation.

### 6.2.6 Cell visualization and counting by CARD-FISH

Microbial analysis of biomass samples collected from the three BTF and from the activity assays was performed by CARD-FISH, as described by Cassarini et al. (2017). For dual-CARD-FISH, peroxidases of initial hybridizations were inactivated according to the procedure described in Holler et al. (2011). Tyramide amplification was performed using the fluorochromes Oregon Green 488-X and Alexa Fluor 594, which were prepared according to the procedure outlined in Pernthaler et al. (2004).

The microorganisms were visualized using archaeal and bacterial HRP-labeled oligonucleotide probes ARCH915 (Stahl and Amann, 1991) and EUB338-I-III (Daims et al., 1999), respectively. The probes DSS658 (Manz et al., 1998) and ANME-2 538 (Schreiber et al., 2010) were used for the detection of DSS and ANME-2 cells, respectively. Oligonucleotide probes were purchased from Biomers (Ulm, Germany).

All the cells were counterstained with 4', 6'-diamidino-2-phenylindole (DAPI) and visualized using an epifluorescence microscope (Carl Zeiss, Germany). 700-1,000 DAPI-stained cells and their corresponding probe fluorescent signals for each probe were considered for cell counting as described previously in the literature (Musat et al., 2008; Siegert et al., 2011; Kleindienst et al., 2012).

### 6.2.7 DNA extraction

DNA was extracted using a FastDNA® SPIN Kit for soil (MP Biomedicals, Solon, OH, USA) following the manufacturer's protocol. Approximately 0.5 g of the sediment was used for DNA extraction from the initial inoculum and ~0.5 ml of liquid obtained by washing the polyurethane foam packing with nuclease free water was used for extracting the DNA from the BTF biomass. The extracted DNA was quantified and its quality was checked according to the procedure outlined by Bhattarai et al. (2017a).

### 6.2.8 Polymerase chain reaction (PCR) amplification for 16S rRNA genes and Illumina Miseq data processing

The DNA was amplified using bar coded archaea specific primer pair Arc516F and reverse Arc855R. The PCR reaction mixture was prepared as described by Bhattarai et al. (2017a), however, the PCR amplification was performed using a touch-down temperature program. PCR conditions consisted of a pre-denaturation step of 5 min at 95°C, followed by 10 touch-down cycles of 95°C for 30 sec, annealing at 68°C for 30 sec with a decrement per cycle to reach the optimized annealing temperature of 63°C and extension at 72°C. This was followed by 25 cycles of denaturation at 95°C for 30 sec and 30 sec of annealing and extension at 72°C. The final elongation step was extended for 10 min.

The primer pairs used for bacteria were forward bac520F 5'-3' AYT GGG YDT AAA GNG and reverse Bac802R 5'-3' TAC NNG GGT ATC TAA TCC (Song et al., 2013). The following program was used: initial denaturation step at 94°C for 5 min, followed by denaturation at 94°C for 40 sec, annealing at 42 °C for 55 sec and elongation at 72°C for 40 sec (30 cycles). The final elongation step was extended to 10 min. 5 μl of the amplicons were visualized by standard agarose gel electrophoresis at the following conditions: 1% agarose gel, a running voltage of 120 V for 30 min, stained by gel red, and documented using a UV transilluminator fitted with a Gel Doc XR System (Bio-Rad, USA).

After checking the correct band size, 150 μl of PCR amplicons were loaded in 1% agarose gel and electrophoresis was performed for 120 min at 120 V. The gel bands were excited under UV light and the PCR amplicons were cleaned using E.Z.N.A.® Gel Extraction Kit by following the manufacturer's protocol (Omega Biotek, USA). The purified DNA amplicons were sequenced by an Illumina HiSeq 2000 (Illumina, San Diego, USA) and analyzed according to the detailed analytical procedure described in Bhattarai et al. (2017a). A total of 40,000 (± 20,000) sequences were assigned to archaea and bacteria by examining the tags assigned to the amplicons. These sequence data have been submitted to the NCBI GenBank database under BioProject accession number PRJNA415004 (direct link: http://www.ncbi.nlm.nih.gov/bioproject/415004).

## 6.3   Results

### 6.3.1   SO$_4^{2-}$ BTF

The pH of the SO$_4^{2-}$ BTF increased from 7.0 to 8.1 during the first 100 days of operation and then the pH remained nearly constant until the end of the experiment (Figure 6.1a). In phase I, sulfate was consumed while sulfide was scarcely produced (0.6 mM, Figure 6.1b). The sulfate consumption and total dissolved sulfide production rates were 0.25 and 0.03 mmol l$^{-1}$ day$^{-1}$, respectively. At the end of phase I, 5.5 mM of sulfur as sulfate was reduced, but only 9.8% was recovered as total dissolved sulfide, which probably precipitated as metal sulfide.

In phase II, the sulfate consumption rate was 0.13 mmol l$^{-1}$ day$^{-1}$, however, similar to phase I, the total dissolved sulfide concentration was low and it started to increase only after 100 days of reactor operation. On day 125, 7.4 mM of sulfate was consumed and 23% of the reduced sulfate was recovered as total dissolved sulfide. On day 130, the color of the mineral medium changed from transparent to greenish-yellow until the seawater medium was replaced on day 142 and the sulfide concentration decreased from 1.8 to 0.04 mM.

In phase III, the total dissolved sulfide concentrations varied between 0.03 and 1.8 mM, while the sulfate concentration decreased at the same rate as observed previously in phase II. In phases IV, V and VI, the total dissolved sulfide concentration increased to values as high as 6 mM. The SR rate ranged between 0.29 and 0.32 mmol l$^{-1}$ day$^{-1}$ during the last three phases of SO$_4^{2-}$ BTF operation. The sulfide production rate was 0.11 mmol l$^{-1}$ day$^{-1}$ in phases IV and VI, while in phase V, the sulfide production rate was the highest at 0.2 mmol l$^{-1}$ day$^{-1}$. In phase V, 74% of sulfur from SO$_4^{2-}$ was recovered as total dissolved sulfide.

The concentration of methane consumed, i.e. the difference between the concentration of methane in the inlet and outlet of the BTF, and carbon dioxide produced, i.e. the difference between the concentration of carbon dioxide in the outlet and inlet, were the highest in the last four phases (Figure 6.1c) when the sulfide production was also the highest (6 mM). However, this amount corresponds to the net methane concentration and does not account for the formation of methane due to possible methanogenic activity or carbon dioxide production from sources other than methane.

Sediment samples were collected and fixed for CARD-FISH analysis three times during the BTF operation of 230 days. On day 84, neither ANME nor DSS were detected with the probes used (Figure 6.2a). On day 155, ANME-2 were detected in low amounts in the fixed samples (Figure 6.2b); however, at the end of the experiment (day 230), ANME-2 cells in the analyzed sediment were distinguishably abundant and cocci-shaped (Figure 6.2c).

### 6.3.2   S$^0$ BTF

The pH of the S$^0$ BTF varied from 4.5 to 8.4, and it was only adjusted whenever the pH dropped below 5.0 (Figure 6.1d), primarily due to sulfate production. Figure 6.1e shows the sulfate and sulfide concentration profiles in the S$^0$ BTF. It was assumed that methane is oxidized to bicarbonate, while four moles of elemental sulfur were reduced to four moles of sulfide according to the stoichiometry shown in Eq. 6.3 (Table 6.1). In phase I, sulfate was produced at a rate of 0.25 mmol $l^{-1}$ day$^{-1}$, while hardly any total dissolved sulfide was formed (0.1 mM). In phase II (49-139 days), after replacing the seawater medium, the sulfate concentration increased again as in phase I at a rate of 0.30 mmol $l^{-1}$ day$^{-1}$, while the maximum total dissolved sulfide formed was only 0.34 mM. The carbon dioxide production was very low and nearly constant during the entire $S^0$ BTF operation (Figure 6.1f).

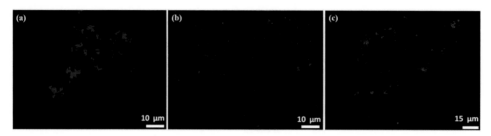

**Figure 6.2** CARD-FISH images of the BTF with sulfate (a) after 84 days of incubation in the reactor showing all living cells stained with DAPI; (b) after 155 days of incubation and (c) after 230 days of incubation showing all living cells stained with DAPI in green and ANME in red (mixture of probes: ANME 1, 2 and 3).

### 6.3.3 Archaeal and bacterial 16S rRNA genes relative abundance

Microbial community profiling was done for the $SO_4^{2-}$ BTF, the $S^0$ BTF and for the BTF with thiosulfate as electron acceptor ($S_2O_3^{2-}$ BTF previously reported by Cassarini et al., 2017). Figure 6.3 shows the results obtained before the inoculation of the three BTFs and at the end of each BTF operation. The highest percentage of archaeal 16S rRNA reads are shown in Figure 6.3a. Among the ANME clades, ANME-2a/b comprised 37% of the archaeal reads in the $SO_4^{2-}$ BTF, while only 3% and 1% of the archaeal reads were retrieved as ANME-2a/b from the $S_2O_3^{2-}$ BTF and $S^0$ BTF, respectively. Other ANME clades, i.e. ANME-1, were also retrieved from the enriched biomass of the three BTF. However, the relative abundance was very low in the case of the $S_2O_3^{2-}$ BTF and $S^0$ BTF. Most of the archaeal reads for the clade ANME-1, specifically the ANME-1b type, were found in the $S_2O_3^{2-}$ BTF (1%).

At the end of each BTF operation, a high percentage of bacterial 16S rRNA reads belonging to the order of *Desulfobacterales* was noticed (Figure 6.3b). In the $SO_4^{2-}$ BTF, a high percentage of *Desulfosarcina* (36%) and SEEP-SRB1 (10%) were retrieved. In the $S_2O_3^{2-}$ BTF, the *Desulfubacterium* (9%) and *Desulfosarcina* sequences were the highest in abundance (5%) within the *Desulfobacterales* order. However, the percentage of the *Sulfurimonas* reads was the highest at 38%. In the $S^0$ BTF, the highest percentage of reads was represented by *Desulfosarcina* (31%) and *Sulfurimonas* (50%) genes, respectively.

### 6.3.4 AOM activity and cell visualization in the $SO_4^{2-}$ BTF

During the activity assays with $^{13}CH_4$ using the biomass from the $SO_4^{2-}$ BTF, 99.5% of the reduced sulfate was recovered as total dissolved sulfide (Figure 6.4a). In the control incubation without biomass (Figure 6.4a, dashed lines), sulfide was not produced, while in the batch incubations without methane (Figure 6.4a, dotted lines), sulfide was produced (1 mM). The DIC produced from methane was calculated and it increased only in the incubations with the biomass from the $SO_4^{2-}$ BTF, in the presence of methane in the headspace (Figure 6.4d). The AOM rate was calculated in the activity assays from the DIC produced from $^{13}CH_4$ (8.4 μmol $l^{-1}$ $day^{-1}$), which was found to be more than 7 times lower than the SR observed in the batch incubations (67.4 mmol $l^{-1}$ $day^{-1}$) (Table 6.2).

The cells retrieved from the $SO_4^{2-}$ BTF after 230 days of operation and from the batch activity assays (35 days) were stained with the general probes for archaea and bacteria (Figure 6.5). The results showed that the bacterial population was more abundant (83%) than the archaeal population (17%). Considering all the stained cells, the ANME-2 cells were less abundant than the DSS cells (7% and 46%, respectively). However, the stained ANME-2 cells constituted 43% of the total amount of stained archaea, while the DSS were 55% of the total amount of stained bacteria.

At the end of the batch activity assays (day 35), the cocci-shaped ANME-2 cells were always visualized in the form of aggregates (Figures 6.5a-6.5c). The DSS cells were present in different shapes, either as cocci (Figures 6.5d and 6.5e) or vibrio-shaped (Figures 6.5e and 6.5f). However, the vibrio-shaped DSS cells were more than three times less abundant than the cocci-shaped DSS cells (23% and 77%, respectively). The aggregates were composed of cocci-shaped DSS and ANME-2 cells (Figures 6.5g and 6.5h). The cocci-shaped DSS cells were more abundant than the ANME-2 cells (Figure 6.5j) and were also visualized without their archaeal partner (Figures 6.5k and 6.5l). Vibrio-shaped DSS cells were always visualized alone, distant from ANME-2 cells (Figure 6.5i).

### 6.3.5    AOM activity and cell visualization in the $S^0$ BTF

During the activity assays with $^{13}CH_4$, the total dissolved sulfide concentration in the incubations with $S^0$ BTF biomass increased from 0 to 2.0 (± 0.4) mM, while the sulfate concentration remained nearly constant (0.9 ± 0.1 mM, Figure 6.4b). In control experiments without the biomass or without methane, sulfide and sulfate were not produced. The DIC produced from $^{13}CH_4$ was calculated and it increased only in the incubations with the biomass from the $S^0$ BTF with methane in the headspace during the first 27 days, but thereafter it decreased (Figure 6.4e). The AOM rate was 6.8 μmol $l^{-1}$ $day^{-1}$. This rate was the lowest among the three BTF (Table 6.2), even more than 7 times lower than the sulfide production rate observed in the batch incubations with $S^0$ BTF biomass (60.7 mmol $l^{-1}$ $day^{-1}$, Table 6.2). Vibrio-shaped DSS cells were abundant in the $S^0$ BTF after 147 days of operation (Figure 6.6a and 6.6b). Few cocci-shaped ANME-2 cells were also visualized in aggregates, either alone or with other unidentified cells stained with DAPI (Figures 6.6c and 6.6d).

### 6.3.6    AOM activity in the $S_2O_3^{2-}$ BTF

During the previously reported $S_2O_3^{2-}$ BTF operation (Cassarini et al., 2017), the disproportionation of $S_2O_3^{2-}$ to $SO_4^{2-}$ and sulfide was the dominant process, with a sulfide production rate of 0.5 mmol $l^{-1}$ $day^{-1}$. Moreover, in the absence of thiosulfate, the sulfate produced was consumed in the presence of methane as the sole electron acceptor (0.38 mmol of sulfate l$^{-1}$ day$^{-1}$).

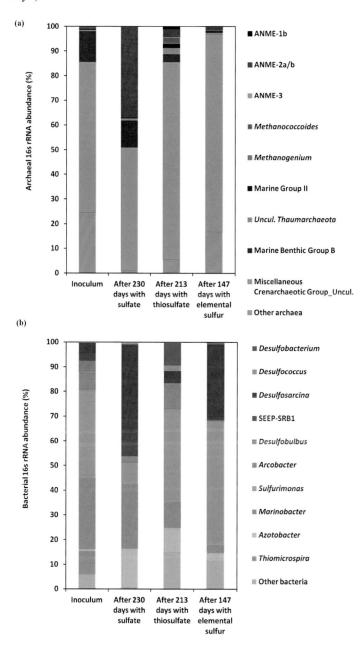

**Figure 6.3** Top most abundant 16S rRNA sequences showing the phylogenetic affiliation up to gene level as derived by high throughput sequencing of archaea (a) and bacteria (b) for the initial inoculum and the inoculum enriched in the SO$_4^{2-}$ BTF (230 days), S$_2$O$_3^{2-}$ BTF (213 days) and S$^0$ BTF (147 days), respectively, at the end of each BTF operation.

During the batch activity assays with $^{13}CH_4$ and thiosulfate as electron acceptor, 99% of the thiosulfate reduced was recovered as total dissolved sulfide and sulfate, respectively (Figure 6.4c). In the control incubations without the biomass, sulfide and sulfate were not produced, while in the incubations without methane, both sulfide and $SO_4^{2-}$ were produced (4 and 3.6 mM, respectively). The DIC produced increased only in the samples with the biomass from the $S_2O_3^{2-}$ BTF and in the presence of methane in the headspace (Figure 6.4f). The AOM rate was 11.5 µmol l$^{-1}$ day$^{-1}$, which was found to be higher than the AOM rate obtained in the $SO_4^{2-}$ BTF (8.4 µmol l$^{-1}$ day$^{-1}$). However, this value was ~10 times lower than the thiosulfate reduction rate (112.6 µmol l$^{-1}$ day$^{-1}$, Table 6.2) determined from the batch activity assays.

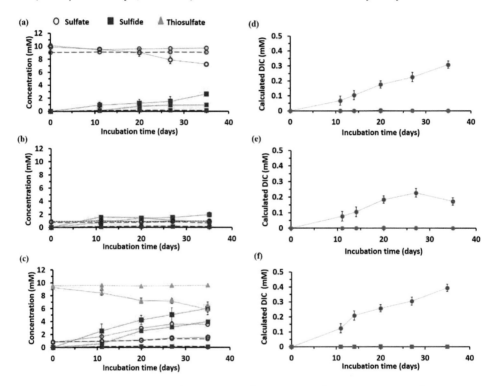

**Figure 6.4** Batch activity assay profiles for the $SO_4^{2-}$ BTF (a, d), $S^0$ BTF (b, e) and $S_2O_3^{2-}$ BTF (c, f). Sulfide and sulfate profiles (a, b, c) and DIC production (d, e, f) during the activity test for the batches incubated with $^{13}CH_4$ (triplicates) and controls. Dotted lines show the controls without methane but nitrogen in the headspace and dashed lines show the controls without biomass. Error bars represent the standard deviation of triplicate measurements.

## 6.4    Discussion

### 6.4.1   Performance of CH₄ oxidizing BTF using different electron acceptors

This study shows that different SR and AOM rates are achieved in identical BTFs when providing the same biomass with different sulfur compounds as electron acceptors for AOM.

In the $SO_4^{2-}$ BTF (Figure 6.1b) and $S_2O_3^{2-}$ BTF (Cassarini et al., 2017), the retardation of the sulfide production compared to the sulfate and thiosulfate reduction might be due to the presence of iron oxides in the inoculum sediment (Wehrmann et al., 2011), favoring the chemical oxidation of sulfide to elemental sulfur and sulfate, respectively, or the precipitation of sulfide as pyrite ($FeS_2$) (Finster et al., 1998; Wan et al., 2014; Cassarini et al., 2017). Moreover, the formation of other sulfur compounds, such as polysulfides or elemental sulfur is also possible, supported by the observed change of color of the mineral medium (Finster et al., 1998) on day 130 in the $SO_4^{2-}$ BTF and on day 18 in the $S_2O_3^{2-}$ BTF (Cassarini et al., 2017).

**Table 6.2** Cumulative rates of SR, sulfide production and anaerobic methane oxidation in batch activity assays with the biomass withdrawn from the three BTF incubated with sulfate, elemental sulfur and thiosulfate.

| Reactor | Sulfate/thiosulfate consumption | Sulfide production | Anaerobic methane oxidation |
|---|---|---|---|
| | | $\mu mol\ l^{-1}\ day^{-1}$ | |
| $SO_4^{2-}$ BTF | 67.4 ($SO_4^{2-}$) | 65.1 | 8.4 |
| $S^0$ BTF | Not determined | 60.7 | 6.8 |
| $S_2O_3^{2-}$ BTF | 112.6 ($S_2O_3^{2-}$) | 185.1 | 11.5 |

At the end of the BTF operation, the $SO_4^{2-}$ BTF followed the stoichiometry of the reaction for AOM-SR (Eq. 6.1, Table 6.1), while the $S_2O_3^{2-}$ BTF followed the stoichiometry for thiosulfate disproportionation (Eq. 6.4, Table 6.1). The sulfide production rate was higher in the $S_2O_3^{2-}$ BTF (0.5 mmol $l^{-1}$ $day^{-1}$) than in the $SO_4^{2-}$ BTF (0.2 mmol $l^{-1}$ $day^{-1}$) due to thiosulfate disproportionation, which is more energetically favorable at the standard operating conditions of the BTF (Eqs. 6.1 and 6.4, Table 6.1). The highest SR rate in the $SO_4^{2-}$ BTF (0.3 mmol $l^{-1}$ $day^{-1}$) was achieved after 142 days of reactor operation. A slightly higher SR rate (0.38 mmol $l^{-1}$ $day^{-1}$) was obtained in the $S_2O_3^{2-}$ BTF only after 46 days of operation. This confirms the hypothesis suggested in our previous study (Cassarini et al., 2017) that the initial addition of thiosulfate decreases the start-up time required for SR in anaerobic methane oxidizing BTF. Sulfide produced by thiosulfate disproportionation is a reducing agent for the seawater medium, which could have possibly accelerated the AOM activity. Thus, addition of a strong reducing agent may also speed up the start-up of SR in anaerobic methane oxidizing reactors. However, the start-up time is a critical step for any bioreactor operation and the long start-up time of the AOM process has been one of the major hindrances for its biotechnological application in in the treatment of groundwater, mine wastewater and wastewaters rich in inorganics (Meulepas et al., 2009a).

The SR rate achieved in the $SO_4^{2-}$ BTF (0.3 mmol $l^{-1}$ $day^{-1}$) was almost half of the highest volumetric rate (0.6 mmol $l^{-1}$ $day^{-1}$) obtained in a membrane bioreactor operated at atmospheric pressure after 884 days of operation (Meulepas et al., 2009a). Long start-up periods, i.e. ~400 days (Meulepas et al., 2009a) and ~365 days (Aoki et al., 2014), have been reported in previous studies where high SR rates were reported. In comparison, in this study, sulfate was reduced almost instantaneously after reactor start-up and the highest SR rate was achieved after 140 days, which shows that the BTF is a good reactor configuration for AOM-SR and the enrichment of microorganisms mediating AOM-SR.

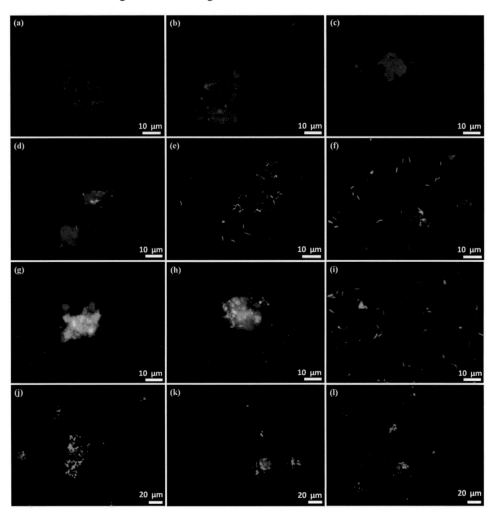

**Figure 6.5** CARD-FISH images for the batch activity assays with sulfate with ANME2-538 probe in red (a, b, c) and with DSS658 probe in green (d, e, f). All cells were counterstained with DAPI in blue. Dual-CARD-FISH with ANME2-538 probe in red and DSS658 probe in green (g-l). The images were taken from the biomass collected and chemically fixed at the end of the batch incubation with $^{13}CH_4$ (35 days).

The BTF reactor configuration is widely used for the treatment of volatile organic and inorganic compounds in waste gases (Santos et al., 2015; Pérez et al., 2016). The BTF reactors used to treat methane emissions under aerobic conditions usually have a short start-up period of ~1-2 weeks (Avalos Ramirez et al., 2012; Estrada et al., 2014a; Estrada et al., 2014b), much lower compared to the observed 42 days ($S_2O_3^{2-}$ BTF, Cassarini et al., 2017) and 140 days ($SO_4^{2-}$ BTF, Figure 6.1b) necessary to obtain the highest SR under anaerobic conditions. The efficient gas-liquid mass transfer and high biomass retention capacity of the BTF technology provide an efficient solution for the enrichment of the slow growing AOM-SR consortia. According to Li et al. (2014), packing materials that offer high porosity, large specific surface area, high robustness, high surface roughness, and moderate grain size should be the preferred choice for microbial attachment and enhanced gas to liquid mass transfer. Besides, the BTF has several advantages over other bioreactor configurations, for instance it is more effective than a submerged (Meulepas et al., 2009a) and external ultrafiltration (Bhattarai et al., 2017 submitted) membrane reactor because it accelerates microbial growth during the start-up period. Moreover, the high investment and operational costs of high pressure bioreactors can be avoided by selecting a BTF operating at ambient temperature and pressure conditions.

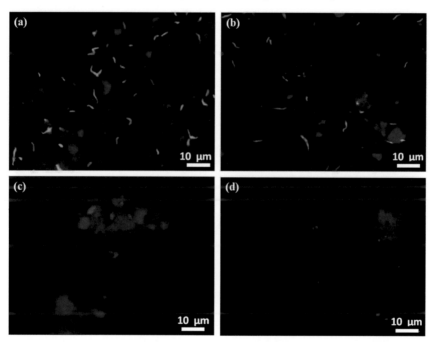

**Figure 6.6** CARD-FISH images for the activity assays with elemental sulfur (a-b) with DSS658 probe (green); and (c-d) with ANME2-538 probe (red). All cells were stained with DAPI (blue). The images were taken from the biomass collected and fixed at the end of the batch incubation with $^{13}CH_4$ (35 days).

In the $S^0$ BTF, no SR occurred; however, on the contrary, sulfate was produced at a maximum rate of 0.3 mmol l$^{-1}$ day$^{-1}$ probably due to elemental sulfur disproportionation (Eq. 6.5, Table 6.1), similar to the observations made in the $S_2O_3^{2-}$ BTF (Cassarini et al., 2017). The disproportionation of $S^0$ requires energy if sulfide accumulates, unless an oxidant such as Fe (III) renders the reaction more energetically favorable by trapping the sulfide (Finster, 2008). Sulfur disproportionation becomes energetically more favorable under alkaline conditions. Therefore, in alkaline environments, such as soda lakes, sulfur disproportionation by haloalkaliphilic bacteria can proceed in the absence of sulfide trapping substances (Poser et al., 2013). The pH and salinity of the $S^0$ BTF was not suitable for the growth of haloalkaliphilic bacteria, but the iron oxides present in the sediment could have acted as sulfide scavenger thereby rendering the microbial disproportionation more favorable. In a recent study, Wegener et al. (2016) showed that neutrophilic sulfur disproportionating bacteria, i.e. *Desulfocapsa sulfoxigens*, are even able to disproportionate sulfur without the addition of iron.

According to the stoichiometry (Eq. 6.5, Table 6.1), sulfate and dissolved sulfide should be produced in a ratio of 1:3, while hardly any dissolved sulfide was produced during the entire $S^0$ BTF operation. Thus, in the $S^0$ BTF, the chemical oxidation of dissolved sulfide by iron oxides does not completely explain the sulfide loss. The $S^0$ BTF was always maintained under anaerobic conditions, thus aerobic oxidation of sulfur and sulfide was highly unlikely. However, the pH decreased during its operation and this pH drop can be explained by the reaction stoichiometry of either elemental sulfur disproportionation (Eq. 6.5, Table 6.1) or elemental sulfur reduction (Eq. 6.3, Table 6.1). Besides, the decrease in pH as well as the increase in sulfide concentration renders both reactions thermodynamically less favorable. The pH should have been maintained at values > 7.0 to facilitate the occurrence of elemental sulfur disproportionation and/or reduction.

### 6.4.2   Rates of AOM and sulfur reduction

The AOM rates were much lower than the sulfate and thiosulfate reduction rates and the sulfide production rates (Table 6.2). In previous studies, it was shown that trace amounts of methane oxidation occurs while net methane production is observed (Meulepas et al., 2010; Timmers et al., 2015b). Therefore, part of the $^{13}C$-DIC could be due to trace methane oxidation during methanogenesis. However, no net SR occurred in the batch incubations with sulfate, in the absence of methane (Figure 6.4a) and methane was never detected in any of the incubations without methane (either with sulfate, thiosulfate or elemental sulfur as electron acceptors), excluding the occurrence of methanogenesis and concomitant trace methane oxidation.

The maximum dissolved methane concentration at a salinity of 32‰ and 20°C is 1.27 mM (Yamamoto et al., 1976). This value was used for the estimation of the $\Delta_r G^{'}$ (Table 6.2). 31% of the methane was converted to carbon dioxide in the batch incubations containing thiosulfate as the electron acceptor, while with sulfate and elemental sulfur only 25% and 10% of methane was consumed. Therefore, the use of thiosulfate as the electron acceptor not only reduces the start-up time required for SR, but it also triggers higher AOM rates (within 213 days of operation) than the sulfate fed BTF that was operated for 230 days.

The volumetric rates of sulfate and thiosulfate consumption and sulfide production obtained in the three BTF were approximately three times higher than the rates obtained during the batch activity assays (Table 6.2). This is probably due to the activity of the inoculum used and the

BTF hydrodynamics wherein the artificial sea water medium was trickled from the top through the packed polyurethane foam cubes that host the inoculum, while the methane was supplied in up-flow mode that enables good gas to liquid contact. This advantageous mode of reactor operation might have facilitated better thiosulfate and sulfate consumption and supported the growth of the slow growing ANME and SRB in the BTF (Figure 6.1). The sulfide production rates obtained in similar studies (Meulepas et al., 2009b; Suarez-Zuluaga et al., 2014) with thiosulfate as the electron acceptor and methane as the sole electron donor were lower (0.086 and 0.11 mmol $l^{-1}$ $day^{-1}$, respectively) than the maximum sulfide production rates obtained during the BTF operation with thiosulfate (0.5 mmol $l^{-1}$ $day^{-1}$) and also the batch activity assays (Table 6.2).

The biomass transferred from the $S_2O_3^{2-}$ BTF after 213 days of enrichment to perform batch activity assays showed a thiosulfate transformation mainly to sulfide and concomitant AOM, suggesting that AOM was directly coupled to thiosulfate reduction. Alternatively, as suggested in previous studies (Meulepas et al., 2009b; Cassarini et al., 2017) a two step process mediated by two different groups of bacteria might have occurred: thiosulfate disproportionating bacteria and sulfate reducing bacteria, the latter scavenging the sulfate produced by the first group. This two-step process could also explain the differences noticed between the thiosulfate consumption and AOM rates (Table 6.2).

### 6.4.3   Microorganisms enriched in the three BTF

The highest AOM rate is expected to occur where anaerobic methanotrophs, i.e. ANME, are more abundant. However, CARD-FISH analysis (Cassarini et al., 2017) showed that in the $S_2O_3^{2-}$ BTF, a high number of vibrio-shaped DSS stained cells were found and they increased from 40% to 70% of the total number of counted cells, but very few cells were stained with the ANME-2 probe (less than 1%). In contrast, in the $SO_4^{2-}$ BTF, 43% of the total archaea cells were stained with the ANME-2 probe, which is also in accordance with the relative abundance of the 16S rRNA sequences (Figure 6.3a). Whereas in the case of the $S_2O_3^{2-}$ BTF, 1% of archaeal sequences were retrieved as ANME-1. This clade not considered in the previously performed CARD-FISH analysis (Cassarini et al., 2017), might have been responsible for AOM in the $S_2O_3^{2-}$ BTF, even if its relative abundance is low.

The DSS probe stained a high number of bacterial cells in the three BTF at the end of the reactors operation, i.e. 55% ($SO_4^{2-}$ BTF), 70% ($S_2O_3^{2-}$ BTF) and 60% ($S^0$ BTF) of the total amount of stained bacteria. The stained DSS cells had a different morphology (vibrio-shaped or coccoid morphology), which suggests a genomic difference of the microorganisms (Schreiber et al., 2010). The DSS658 horseradish peroxidase (HRP)-labeled probe not only targets the *Desulfosarcina/Desulfococcus* group, but also targets groups from the *Desulfobacteraceae* family and other groups of SRB. 16S rRNA sequences from the *Desulfobacteraceae* family were retrieved from the three BTF (Figure 6.3b) and therefore the stained DSS cells could also belong to other genera of that family.

A high abundance of vibrio-shaped DSS cells was visualized in the $S_2O_3^{2-}$ BTF and $S^0$ BTF, respectively, while cocci-shaped DSS cells were more abundant in the $SO_4^{2-}$ BTF. Members of the SEEP-SRB1 (a clade of SRB), the *Desulfobacterium* and *Desulfosarcina* groups, were reported to have a vibrio- or rod-shaped morphology (Castro et al., 2000; Schreiber et al., 2010; Kuever et al., 2015). Species from these groups capable of reducing sulfate and other oxidized sulfur compounds to sulfide might be the microorganisms enriched in the $S_2O_3^{2-}$ BTF and $S^0$ BTF, respectively. However, in these two BTF, sulfur disproportionation prevailed, but none of the sequences was related to the most commonly reported thiosulfate and elemental sulfur disproportionating bacteria, the vibrio-shaped *Desulfovibrio* group (Castro et al., 2000) or the *Desulfocapsa* group (Finster, 2008; Suarez-Zuluaga et al., 2014).

The only bacterial sequences found in this study that have been previously reported as thiosulfate or elemental sulfur disproportionating bacteria are the *Desulfobulbus* relatives (Finster, 2008; Pjevac et al., 2014). However, they were relatively low in abundance (Figure 6.3b). A high percentage of bacterial reads related to the *Epsilonproteobacteria* class, i.e. *Sulfurimonas* and *Arcobacter*, was found in the original sediment and in similar percentage in the $S^0$ and $S_2O_3^{2-}$ enriched BTF, while they were less abundant in the $SO_4^{2-}$ BTF. These microorganisms have vibrio-shaped and filamentous morphology (Schauer et al., 2011), but are not be stained with the DSS658 probe. These sulfide oxidizers are usually found in anaerobic environments as nitrate or nitrite reducers (Grote et al., 2012; Pjevac et al., 2014; Han and Perner, 2015). They could have been responsible for some sulfide oxidation, especially during the start of the three BTF, when the total dissolved sulfide was hardly detectable. It is noteworthy to mention that nitrate or nitrite was not added to the initial seawater medium. Alternatively, *Sulfurimonas* might have been involved in the disproportionation of $S_2O_3^{2-}$ and $S^0$.

This study showed that the BTF is a suitable reactor configuration for the enrichment of slow growing microorganisms using either sulfate or thiosulfate as the electron acceptor. The use of thiosulfate triggered the highest AOM rate, but it is still unclear which microorganisms are the key players involved in methane oxidation and further investigation on the role and the metabolic activity of the microorganisms involved in AOM coupled to thiosulfate is required, e.g. using labeled substrates and nanometre scale secondary ion mass spectrometry (NanoSIMS) (Musat et al., 2016).

The microbial analysis from the enriched biomass in the $S_2O_3^{2-}$ BTF confirmed that thiosulfate as electron acceptor triggered the growth of DSS. Nevertheless, in order to enhance the growth of ANME and avoid sulfate production by disproportionation, sulfate should be used as the electron acceptor. Sulfate was the preferred electron acceptor for the growth of ANME-2 cells under ambient conditions and by the sediment used as inoculum in this study (Figure 6.5). From a practical viewpoint, thiosulfate and sulfate can be used as electron acceptors to accelerate the growth of SRB and ANME in a BTF when AOM-SR is used for the simultaneous removal of methane and oxidized sulfur compounds from waste streams.

## 6.5   References

Acree, T.E., Sonoff, E.P., Splittstoesser, D.F., 1971. Determination of hydrogen sulfide in fermentation broths containing $SO_2$. *Appl. Microbiol.*, **22**(1), 110-112.

Aoki, M., Ehara, M., Saito, Y., Yoshioka, H., Miyazaki, M., Saito, Y., Miyashita, A., Kawakami, S., Yamaguchi, T., Ohashi, A., Nunoura, T., Takai, K., Imachi, H. 2014. A long-term cultivation of an anaerobic methane-oxidizing microbial community from deep-sea methane-seep sediment using a continuous-flow bioreactor. *PLoS ONE*, **9**(8), Pe105356.

APHA. 1995. Standard methods for the examination of water and wastewater. *American Public Health Association*, (19th edition), Washington DC, USA. pp.1325.

Avalos Ramirez, A., Jones, J.P., Heitz, M. 2012. Methane treatment in biotrickling filters packed with inert materials in presence of a non-ionic surfactant. *J. Chem. Technol. Biotechnol.*, **87**(6), 848-853.

Bhattarai, S., Cassarini, C., Gonzalez-Gil, G., Egger, M., Slomp, C.P., Zhang, Y., Esposito, G., Lens, P.N.L. 2017a. Anaerobic methane-oxidizing microbial community in a coastal marine sediment: anaerobic methanotrophy dominated by ANME-3. *Microb. Ecol.* **74**(3), 608-622.

Bhattarai, S., Cassarini, C., Naangmenyele, Z., Rene, E. R., Gonzalez-Gil, G., Esposito, G., Lens, P.N.L. 2017b. Microbial sulfate reducing activities in anoxic sediment from marine lake Grevelingen. *Limnology*, **19**(1), 31-41.

Cassarini C., Rene E. R., Bhattarai S., Esposito G., Lens P.N.L. (2017) Anaerobic oxidation of methane coupled to thisoulfate reduction in a biotrickling filter. *Bioresour. Technol.*, **240**(3), 214-222.

Castro, H., Williams, N., Ogram, A. 2000. Phylogeny of sulfate-reducing bacteria. *FEMS Microbiol. Ecol.*, **31**(1), 1-9.

Craig, H. 1957. Isotopic standards for carbon and oxygen and correction factors for mass-spectrometric analysis of carbon dioxide. *Geochim. Cosmochim. Ac.*, **12**(1-2), 133-149.

Daims, H., Brühl, A., Amann, R., Schleifer, K.-H., Wagner, M. 1999. The domain-specific probe EUB338 is insufficient for the detection of all Bacteria: development and evaluation of a more comprehensive probe set. *Syst. Appl. Microbiol.*, **22**(3), 434-444.

Deusner, C., Meyer, V., Ferdelman, T. 2009. High-pressure systems for gas-phase free continuous incubation of enriched marine microbial communities performing anaerobic oxidation of methane. *Biotechnol. Bioeng.*, **105**(3), 524-533.

Dorer, C., Vogt, C., Neu, T.R., Stryhanyuk, H., Richnow, H.-H. 2016. Characterization of toluene and ethylbenzene biodegradation under nitrate-, iron (III)-and manganese (IV)-reducing conditions by compound-specific isotope analysis. *Environ. Pollut.*, **211**(29), 271-281.

Estrada, J.M., Dudek, A., Muñoz, R., Quijano, G. 2014a. Fundamental study on gas-liquid mass transfer in a biotrickling filter packed with polyurethane foam. *J. Chem. Technol. Biotechnol.*, **89**(9), 1419-1424.

Estrada, J.M., Lebrero, R., Quijano, G., Pérez, R., Figueroa-González, I., García-Encina, P.A., Muñoz, R., 2014b. Methane abatement in a gas-recycling biotrickling filter: evaluating innovative operational strategies to overcome mass transfer limitations. *Chem. Eng. J.*, **253**(53), 385-393.

Finster, K. 2008. Microbiological disproportionation of inorganic sulfur compounds. *J. Sulfur Chem.*, **29**(3-4), 281-292.

Finster, K., Liesack, W., Thamdrup, B. 1998. Elemental sulfur and thiosulfate disproportionation by *Desulfocapsa sulfoexigens* sp . nov., a new anaerobic bacterium isolated from marine surface sediment. *Appl. Environ. Microb.*, **64**(1), 119-125.

Forster, A., Schouten, S., Baas, M., Sinninghe Damsté, J.S., 2007. Mid-cretaceous (Albanian-Santonian) sea surface temperature record of the tropical. *Atlantic Ocean. Geology.* **35**(10), 919-919.

Girguis, P.R., Cozen, A.E., DeLong, E.F. 2005. Growth and population dynamics of anaerobic methane-oxidizing archaea and sulfate-reducing bacteria in a continuous-flow bioreactor. *Appl. Environ. Microbiol.*, **71**(7), 3725-3733.

Grote, J., Schott, T., Bruckner, C.G., Glöckner, F.O., Jost, G., Teeling, H., Labrenz, M., Jürgens, K. 2012. Genome and physiology of a model Epsilonproteobacterium responsible for sulfide detoxification in marine oxygen depletion zones. *Proc. Natl. Acad. Sci. USA*, **109**(2), 506-510.

Han, Y., Perner, M. 2015. The globally widespread genus *Sulfurimonas*: versatile energy metabolisms and adaptations to redox clines. *Front. Microbiol.*, **6**(989), 1-17.

Hinrichs, K.-U., Boetius, A. 2002. The anaerobic oxidation of methane: new insights in microbial ecology and biogeochemistry. in: Wefer, G., Billett, D., Hebbeln, D., Jørgensen, B.B., Schlüter, M., van Weering, T.E. (Eds), *Ocean Margin Systems*, Springer Berlin Heidelberg, Germany, pp. 457-477.

Hinrichs, K.-U., Hayes, J.M., Sylva, S.P., Brewer, P.G., DeLong, E.F. 1999. Methane-consuming archaebacteria in marine sediments. *Nature*, **398**(6730), 802-805.

Holler, T., Wegener, G., Niemann, H., Deusner, C., Ferdelman, T.G., Boetius, A., Brunner, B., Widdel, F. 2011. Carbon and sulfur back flux during anaerobic microbial oxidation of methane and coupled sulfate reduction. *Proc. Natl. Acad. Sci. USA*, **108**(52), E1484-E1490.

Kirschke, S., Bousquet, P., Ciais, P., Saunois, M., Canadell, J.G., Dlugokencky, E.J., Bergamaschi, P., Bergmann, D., Blake, D.R., Bruhwiler, L., Cameron-Smith, P.,

Castaldi, S., Chevallier, F., Feng, L., Fraser, A., Heimann, M., Hodson, E.L., Houweling, S., Josse, B., Fraser, P.J., Krummel, P.B., Lamarque, J.-F., Langenfelds, R.L., Le Quere, C., Naik, V., O'Doherty, S., Palmer, P.I., Pison, I., Plummer, D., Poulter, B., Prinn, R.G., Rigby, M., Ringeval, B., Santini, M., Schmidt, M., Shindell, D.T., Simpson, I.J., Spahni, R., Steele, L.P., Strode, S.A., Sudo, K., Szopa, S., van der Werf, G.R., Voulgarakis, A., van Weele, M., Weiss, R.F., Williams, J.E., Zeng, G. 2013. Three decades of global methane sources and sinks. *Nat. Geosci.*, **6**(10), 813-823.

Kleindienst, S., Ramette, A., Amann, R., Knittel, K. 2012. Distribution and *in situ* abundance of sulfate-reducing bacteria in diverse marine hydrocarbon seep sediments. *Environ. Microbiol.*, **14**(10), 2689-2710.

Knittel, K., Boetius, A. 2009. Anaerobic oxidation of methane: progress with an unknown process. *Annu. Rev. Microbiol.*, **63**(1), 311-334.

Krüger, M., Blumenberg, M., Kasten, S., Wieland, A., Känel, L., Klock, J.-H., Michaelis, W., Seifert, R. 2008. A novel, multi-layered methanotrophic microbial mat system growing on the sediment of the Black Sea. *Environ. Microbiol.*, **10**(8), 1934-1947.

Kuever, J., Rainey, F.A., Widdel, F. 2015. *Desulfobacterium*. in: Whitman WB (Ed.), Bergey's *Manual of Systematics of Archaea and Bacteria* doi:10.1002/9781118960608.gbm01012. John Wiley & Sons, Inc., Athens, USA.

Li, Z.-X., Yang, B.-R., Jin, J.-X., Pu, Y.-C., Ding, C. 2014. The operating performance of a biotrickling filter with *Lysinibacillus fusiformis* for the removal of high-loading gaseous chlorobenzene. *Biotechnol. Lett.*, **36**(10), 1971-1979.

Manz, W., Eisenbrecher, M., Neu, T.R., Szewzyk, U., 1998. Abundance and spatial organization of gram-negative sulfate-reducing bacteria in activated sludge investigated by *in situ* probing with specific 16S rRNA targeted oligonucleotides. *FEMS Microbiol. Ecol.*, **25**(1), 43-61.

Meulepas, R.J.W., Jagersma, C.G., Gieteling, J., Buisman, C.J.N., Stams, A.J.M., Lens, P.N.L. 2009a. Enrichment of anaerobic methanotrophs in sulfate-reducing membrane bioreactors. *Biotechnol. Bioeng.*, **104**(3), 458-470.

Meulepas, R.J.W., Jagersma, C.G., Khadem, A.F., Buisman, C.J.N., Stams, A.J.M., Lens, P.N.L. 2009b. Effect of environmental conditions on sulfate reduction with methane as electron donor by an Eckernförde Bay enrichment. *Environ. Sci. Technol.*, **43**(17), 6553-6559.

Meulepas, R.J.W., Jagersma, C.G., Zhang, Y., Petrillo, M., Cai, H., Buisman, C.J.N., Stams, A.J.W., Lens, P.N.L. 2010. Trace methane oxidation and the methane dependency of sulfate reduction in anaerobic granular sludge. *FEMS Microbiol. Ecol.*, **72**(2), 261-271.

Milucka, J., Ferdelman, T.G., Polerecky, L., Franzke, D., Wegener, G., Schmid, M., Lieberwirth, I., Wagner, M., Widdel, F., Kuypers, M.M.M. 2012. Zero-valent sulphur is a key intermediate in marine methane oxidation. *Nature*, **491**(7425), 541-546.

Musat, N., Halm, H., Winterholler, B., Hoppe, P., Peduzzi, S., Hillion, F., Horreard, F., Amann, R., Jørgensen, B.B., Kuypers, M.M.M. 2008. A single-cell view on the ecophysiology of anaerobic phototrophic bacteria. *Proc. Natl. Acad. Sci. USA*, **105**(46), 17861-17866.

Musat, N., Musat, F., Weber, P.K., Pett-Ridge, J. 2016. Tracking microbial interactions with NanoSIMS. *Curr. Opin. Biotech.*, **41**(7), 114-121.

Nauhaus, K., Albrecht, M., Elvert, M., Boetius, A., Widdel, F. 2007. *In vitro* cell growth of marine archaeal-bacterial consortia during anaerobic oxidation of methane with sulfate. *Environ. Microbiol.*, **9**(1), 187-196

Niemann, H., Duarte, J., Hensen, C., Omoregie, E., Magalhaes, V.H., Elvert, M., Pinheiro, L.M., Kopf, A., Boetius, A. 2006. Microbial methane turnover at mud volcanoes of the Gulf of Cadiz. *Geochim. Cosmochim. Ac.*, **70**(21), 5336-5355.

Orphan, V.J., Hinrichs, K.-U., Ussler, W., Paull, C.K., Taylor, L.T., Sylva, S.P., Hayes, J.M., DeLong, E.F. 2001. Comparative analysis of methane-oxidizing archaea and sulfate-reducing bacteria in anoxic marine sediments. *Appl. Environ. Microbiol.*, **67**(4), 1922-1934.

Pérez, M., Álvarez-Hornos, F., Engesser, K., Dobslaw, D., Gabaldón, C. 2016. Removal of 2-butoxyethanol gaseous emissions by biotrickling filtration packed with polyurethane foam. *New Biotechnol.*, **33**(2), 263-272.

Pernthaler Annelie , P.J., Amann Rudolf. 2002. Fluorescence *in situ* hybridization and catalyzed reporter deposition for the identification of marine bacteria. *Appl. Environ. Microbiol.*, **68**(6), 3094-3101.

Pernthaler, A., Pernthaler, J., Amann, R., 2004. Sensitive multi-color fluorescence *in situ* hybridization for the identification of environmental microorganisms, in: Kowalchuk, G., de Bruijn, F., Head, I., Akkermans, A., van Elsas, J.D. (Eds.), *Molecular Microbial Ecology Manual.* Springer, Dordrecht, pp. 711-726.

Pjevac, P., Kamyshny, A., Dyksma, S., Mußmann, M. 2014. Microbial consumption of zero-valence sulfur in marine benthic habitats. *Environ. Microbiol.*, **16**(11), 3416-3430

Poser, A., Lohmayer, R., Vogt, C., Knoeller, K., Planer-Friedrich, B., Sorokin, D., Richnow, H.H., Finster, K. 2013. Disproportionation of elemental sulfur by haloalkiphilic bacteria from soda lakes. *Extremophiles* **17**(6), 1003-1012.

Pruesse, E., Peplies, J., Glöckner, F.O. 2012. SINA: accurate high-throughput multiple sequence alignment of ribosomal RNA genes. *Bioinformatics*, **28**(14), 1823-1829.

Raghoebarsing, A.A., Pol, A., van de Pas-Schoonen, K.T., Smolders, A.J.P., Ettwig, K.F., Rijpstra, W.I.C., Schouten, S., Damsté, J.S.S., Op den Camp, H.J.M., Jetten, M.S.M., Strous, M. 2006. A microbial consortium couples anaerobic methane oxidation to denitrification. *Nature*, **440**(7086), 918-921.

Reeburgh, W.S. 2007. Oceanic methane biogeochemistry. *Chem. Rev.*, **107**(2), 486-513

Santos, A., Guimerà, X., Dorado, A.D., Gamisans, X., Gabriel, D. 2015. Conversion of chemical scrubbers to biotrickling filters for VOCs and H$_2$S treatment at low contact times. *Appl. Microbiol. Biotechnol.*, **99**(1), 67-76.

Schauer, R., Røy, H., Augustin, N., Gennerich, H.-H., Peters, M., Wenzhoefer, F., Amann, R., Meyerdierks, A. 2011. Bacterial sulfur cycling shapes microbial communities in surface sediments of an ultramafic hydrothermal vent field. *Environ. Microbiol.*, **13**(10), 2633-2648.

Scheller, S., Yu, H., Chadwick, G.L., McGlynn, S.E., Orphan, V.J. 2016. Artificial electron acceptors decouple archaeal methane oxidation from sulfate reduction. *Science*, **351**(6274), 703-707.

Schloss, P.D., Westcott, S.L. 2011. Assessing and improving methods used in operational taxonomic unit-based approaches for 16S rRNA gene sequence analysis. *Appl. Environ. Microbiol.*, **77**(10), 3219-3226.

Schreiber, L., Holler, T., Knittel, K., Meyerdierks, A., Amann, R. 2010. Identification of the dominant sulfate-reducing bacterial partner of anaerobic methanotrophs of the ANME-2 clade. *Environ. Microbiol.*, **12**(8), 2327-2340.

Siegert, M., Krüger, M., Teichert, B., Wiedicke, M., Schippers, A. 2011. Anaerobic oxidation of methane at a marine methane seep in a forearc sediment basin off Sumatra, Indian Ocean. *Front. Microbiol.*. **2**(249).

Song, Z.-Q., Wang, F.-P., Zhi, X.-Y., Chen, J.-Q., Zhou, E.-M., Liang, F., Xiao, X., Tang, S.-K., Jiang, H.-C., Zhang, C.L., Dong, H., Li, W.-J. 2013. Bacterial and archaeal diversities in Yunnan and Tibetan hot springs, China. *Environ. Microbiol.*, **15**(4), 1160-1175.

Stahl, D.A., Amann, R.I. 1991. Development and application of nucleic acid probes. In: Stackebrandt, E., Goodfellow, M. (Eds.), *Nucleic acid techniques in bacterial systematics*. John Wiley & Sons Ltd, Chichester, UK, pp. 205-248.

Suarez-Zuluaga, D.A., Timmers, P.H.A., Plugge, C.M., Stams, A.J.M., Buisman, C.J.N., Weijma, J. 2015. Thiosulphate conversion in a methane and acetate fed membrane bioreactor. *Environ. Sci. Pollut. Res.*, **23**(3), 2467-2478.

Suarez-Zuluaga, D.A., Weijma, J., Timmers, P.H.A., Buisman, C.J.N. 2014. High rates of anaerobic oxidation of methane, ethane and propane coupled to thiosulphate reduction. *Environ. Sci. Pollut. Res.*, **22**(5), 3697-3704.

Timmers, P.H., Gieteling, J., Widjaja-Greefkes, H.A., Plugge, C.M., Stams, A.J., Lens, P.N.L., Meulepas, R.J. 2015a. Growth of anaerobic methane-oxidizing archaea and sulfate-reducing bacteria in a high-pressure membrane capsule bioreactor. *Appl. Environ. Microbiol.*, **81**(4), 1286-1296.

Timmers, P.H., Suarez-Zuluaga, D.A., van Rossem, M., Diender, M., Stams, A.J., Plugge, C.M. 2015b. Anaerobic oxidation of methane associated with sulfate reduction in a natural freshwater gas source. *ISME J.*, **10**(6), 1400-1412.

Wan, M., Shchukarev, A., Lohmayer, R., Planer-Friedrich, B., Peiffer, S. 2014. Occurrence of surface polysulfides during the interaction between ferric (hydr)oxides and aqueous sulfide. *Environ. Sci. Technol.*, **48**(9), 5076-5084.

Wegener, G., Krukenberg, V., Ruff, S.E., Kellermann, M.Y., Knittel, K. 2016. Metabolic capabilities of microorganisms involved in and associated with the anaerobic oxidation of methane. *Front. Microbiol.*, **7**(46), 1-16.

Wehrmann, L.M., Risgaard-Petersen, N., Schrum, H.N., Walsh, E.A., Huh, Y., Ikehara, M., Pierre, C., D'Hondt, S., Ferdelman, T.G., Ravelo, A.C., Takahashi, K., Zarikian, C., 2011. Coupled organic and inorganic carbon cycling in the deep subseafloor sediment of the north-eastern bering sea slope. *Chem. Geol.*, **284**(3-4), 251-261.

Wendeberg, A., 2010. Fluorescence *in situ* hybridization for the identification of environmental microbes. *Cold Spring Harb. Protoc.*, **5**(1), 1-9.

Widdel, F., Bak, F. 1992. Gram negative mesophilic sulfate reducing bacteria. in: Balows, A., Truper, H., Dworkin, M., Harder, W., Schleifer, K. H. (Eds.), *The prokaryotes: a handbook on the biology of bacteria: ecophysiology, isolation, identification, applications.*, Vol. 2, Springer New York, USA, pp. 3352-3378.

Yamamoto, S., Alcauskas, J.B., Crozier, T.E. 1976. Solubility of methane in distilled water and seawater. *J. Chem. Eng. Data*, **21**(1), 78-80.

Zhang, Y., Maignien, L., Zhao, X., Wang, F., Boon, N. 2011. Enrichment of a microbial community performing anaerobic oxidation of methane in a continuous high-pressure bioreactor. *BMC Microbiol.*, **11**(137), 1-8.

# CHAPTER 7

## General Discussion and Future Perspectives

## 7.1   Introduction

Anaerobic methane oxidation (AOM) coupled to sulfate reduction (SR) is a known biological process mediated by anaerobic methanotrophic arachea (ANME) and sulfate reducing bacteria (SRB). It occurs in anaerobic environments and it is responsible for the attenuation of the emission of the green-house gas methane ($CH_4$) to the atmosphere. AOM coupled to sulfate reduction (AOM-SR) has potential application in environmental biotechnology as a process for $CH_4$ removal and desulfurization of industrial wastewater. Both *in situ* and *in vitro* studies have been conducted to understand this microbial mediated bioprocess and the cooperative/synergistic mechanisms involved. Despite several detailed researches on this topic, it is still difficult to enrich these slow growing microorganisms and the highest AOM-SR rates reported so far in the literature are ~100 times lower than what is required to apply AOM-SR for biological wastewater treatment. This research investigated new approaches to control AOM-SR and enrich ANME and SRB in a bioreactor. The current knowledge on AOM-SR and the different AOM communities involved have been overviewed in Chapter 2. Evidently, there are several factors affecting the AOM-SR mechanism and rates: the origin of the marine sediment and the type of ANME and SRB involved, the $CH_4$ availability, the substrates available and used by the microorganisms and the way the microorganisms were enriched *in vitro* (e.g. in batch incubation, membrane bioreactor or high-pressure bioreactor). All these factors were taken into account in this PhD research for designing a suitable bioreactor for AOM-SR that was able to operate at ambient pressure and temperature. The major findings from individual chapters of this PhD are shown in Figure 7.1.

## 7.2   AOM community steered by pressure and substrates used

Marine sediments from two different locations were used in this research: sediment from the marine Lake Grevelingen, the Netherlands, collected at a water depth of 45 m (Chapter 3 and 4) and marine sediment from the Alpha Mound in Gulf of Cadiz in Spain, collected at a water depth of 528 m (Chapters 5 and 6). The microorganisms populating these sediments are different (e.g. ANME-2 type more abundant in Alpha Mound sediment, ANME-3 type more abundant in Grevelingen sediment) and they are subjected to different pressures (~0.45 and ~5.3 MPa, respectively) due to the water depth difference.

Several previous research studies have attempted to mimic the environmental *in situ* conditions to enrich the slow growing ANME and SRB (doubling time of 2-7 months). These microorganisms have been frequently found in deep sediments, where the pressure and temperature are far from ambient conditions (e.g. Gulf of Cadiz sediment subjected to pressure higher than 10 MPa and temperature lower than 10°C) and therefore, such environmental conditions are difficult to simulate in the laboratory. The occurrence of AOM-SR in shallow coastal sediments, such as the marine Lake Grevelingen, where pressure and temperature are closer to ambient conditions (0.45 MPa and 15 °C), is therefore appealing. The marine Lake Grevelingen sediment showed capability of both AOM and SR (Chapter 3). The SR was found to be stimulated by the use of a more favorable electron donor, as ethanol; however, SR coupled to $CH_4$ oxidation occurred as well. In contrast, AOM coupled to thiosulfate and elemental sulfur reduction could not be proven and the disproportionation of these two sulfur compounds prevailed in these experiments.

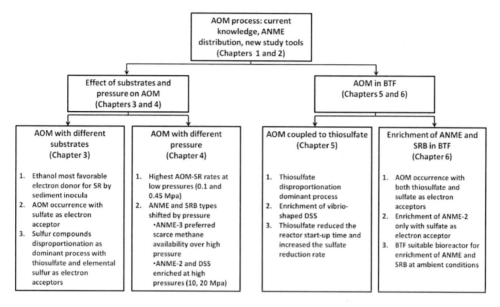

**Figure 7.1** Summary of the major findings of this PhD research

**Note:** AOM - anaerobic oxidation of $CH_4$, SR - sulfate reduction, BTF - biotrickling filter, ANME - anaerobic methanotrophs, SRB - sulfate reducing bacteria and DSS - *Desulfosarcinales/Desulfococcus* group.

The marine Lake Grevelingen hosts both ANME and SRB and among the classic ANME types, ANME-3 was reported to be predominant (Bhattarai et al., 2017). ANME-3 were often found in cold seep areas and mud volcanoes with high $CH_4$ partial pressures and relatively low temperatures (< 20°C) (Losekann et al., 2007; Niemann et al., 2006b; Vigneron et al., 2013). This shallow sediment was a beneficial inoculum to ascertain the pressure effects on ANME-3. Theoretically, elevated $CH_4$ partial pressure favors AOM-SR (Figure 7.2b), as more $CH_4$ will be dissolved and hence it is also bioavailable for the microorganisms. Moreover, previous studies showed strong positive correlation between the growth of ANME and the $CH_4$ partial pressure (up to 12 MPa) (Deusner et al., 2009; Krüger et al., 2005; Nauhaus et al., 2002; Zhang et al., 2010).

Therefore, in this study, the marine Lake Grevelingen sediment was subjected to different $CH_4$ partial pressures (Chapter 4). Surprisingly, the highest AOM-SR activity was obtained at low pressure (0.45 MPa, Figure 7.2d), showing that the active ANME preferred scarce $CH_4$ availability over high pressures (10, 20, 40 Mpa). Interestingly, the abundance and structure of the different type of ANME and SRB were steered by pressure and the ANME-3 type was predominantly enriched at low pressure (Figure 7.2d). Therefore, enriching the ANME and SRB at ambient or close to ambient conditions is feasible by choosing an active AOM inoculum from a shallow sediment or from a sediment rich in ANME and SRB preferring low methane partial pressure, i.e. ANME-3. However, the sediment from Gulf of Cadiz was chosen as inoculum for the enrichment of ANME and SRB in biotrickling filter (BTF) in this research (Chapter 5 and 6), since it is a well known habitat for ANME and SRB (Niemann et al., 2006a; Templer et al., 2011).

AOM coupled to different sulfur compounds as electron acceptor was investigated using sediments from the marine Lake Grevelingen (Chapter 3) and Gulf of Cadiz (Chapters 4 and 5). Sulfate, thiosulfate and elemental sulfur were used as alternative sulfur compounds and electron acceptors. As depicted in Figure 7.2c, thiosulfate as electron acceptor for AOM is theoretically more favorable ($\Delta G^0$= -38.5 kJ mol$^{-1}$ CH$_4$) than sulfate ($\Delta G^0$= -16.6 kJ mol$^{-1}$ CH$_4$). On the other hand, even though elemental sulfur is less favorable ($\Delta G^0$= +24.3 kJ mol$^{-1}$ CH$_4$) than sulfate, it was shown to be directly taken up by ANME (Milucka et al., 2012).

In Chapter 3, it was shown that elemental sulfur and thiosulfate disproportionation to sulfate and sulfide prevailed over their reduction, presumably because disproportionating SRB such as *Desulfocapsa* were enriched (Suarez-Zuluaga et al., 2014). In Chapters 5 and 6, further investigations on the effect of thiosulfate and elemental sulfur on the AOM process were conducted in BTFs. When thiosulfate was used as the electron acceptor, its disproportionation to sulfate and sulfide was the dominating process for sulfur conversion (Chapter 5). However, AOM occurred (Chapter 6) and the enriched SRB belong to the *Desulfosarcinales/Desulfococcus* group (DSS) (Chapter 5 and 6). The biomass enriched in the three BTF with different electron acceptors was used for activity assay incubations to determine the AOM rates. The highest AOM rate was registered using thiosulfate as the electron acceptor (11.5 µmol l$^{-1}$ day$^{-1}$), showing that AOM can be either directly coupled to the reduction of thiosulfate (112.6 µmol l$^{-1}$ day$^{-1}$), or it is a two step process in which AOM is coupled to the reduction of sulfate produced by thiosulfate disproportionation (Figure 7.2e). Moreover, the use of thiosulfate triggered the enrichment of DSS (Figure 7.2e), which are frequently found in association with ANME-2 (Schreiber et al., 2010), while sequences from known disproportionating SRB, such as *Desulfocapsa* or *Desulfovibrio* (Finster, 2008) were not found. Interestingly, hardly any ANME cells could be visualized when thiosulfate was used as electron acceptor and the highest enrichment of ANME-2 was obtained when sulfate was used as the sole electron acceptor (Chapter 6). Further investigation is needed to identify the carbon sources and to quantify the carbon and sulfur fluxes within these enriched microorganisms obtained from the three BTF with different sulfur compounds as electron acceptors.

## 7.3    FISH-NanoSIMS analysis: investigation on AOM-SR and the microorganisms involved

In order to better understand the mechanism of AOM-SR and quantify the metabolic activities at the single-cell level, several approaches have been described in the literature. Among them, microautoradiography, Raman microspectroscopy and nanometre scale secondary ion mass spectrometry (NanoSIMS) are the most widely used analytical techniques (Musat et al., 2012). Combining these analysis to other techniques such as stable isotope probing (SIP) and/or FISH can be used to link the identity, function and metabolic activity at the cellular level to show the metabolic interactions within the consortia (Musat et al., 2016).

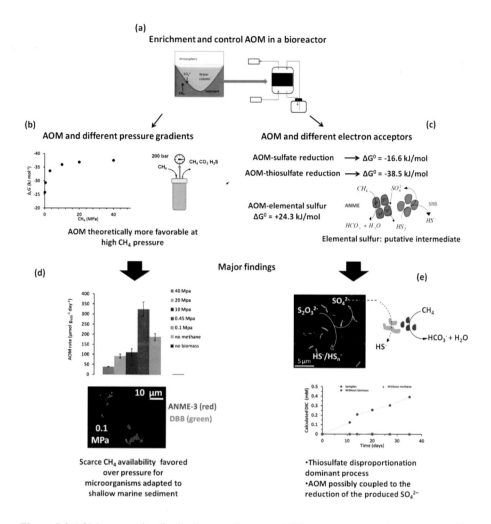

**Figure 7.2** AOM community distribution as a function of different pressure and substrate conditions: hypothesis (b - c) and major findings in this thesis (d - e). (a) The purpose of the study is to control a natural phenomenon (AOM-SR) in a bioreactor; (b) the theoretical influence of the $CH_4$ partial pressure on the Gibbs free energy ($\Delta_r G$ in kJ mol$^{-1}$) of AOM-SR; (c) the standard Gibbs free energy ($\Delta G^0$ in kJ mol$^{-1}$) of the reactions of AOM coupled to different sulfur compounds as electron acceptors and the mechanism proposed by Milucka et al. (2012); (d) major findings from Chapter 4 showing the effect of pressure on AOM rates and CARD-FISH image depicting enriched microorganisms at ambient pressure (0.1 MPa); and (e) major findings of Chapters 5 and 6 showing the putative mechanism and the AOM occurrence with thiosulfate as electron acceptor, confirmed by the production of dissolved inorganic carbon (DIC) from $CH_4$

**Note:** AOM - anaerobic oxidation of $CH_4$, DIC-dissolved inorganic carbon, ANME - anaerobic methanotrophs, SRB - sulfate reducing bacteria and DBB - *Desulfobulbus* group.

In this thesis, FISH-NanoSIMS was used to identify the carbon sources and to quantify the carbon and sulfur fluxes within the microorganisms in the biomass enriched using sulfate as the electron acceptor in a BTF (Chapter 6). In the following paragraphs, the methodology involved in sample preparation for NanoSIMS together with data acquisition and analysis is briefly introduced and the preliminary results obtained will be discussed.

The biomass and the polyurethane foam cubes were transferred from the BTF using sulfate as electron acceptor (Chapter 6) to a 5 l BTF made of glass, which was operated for 180 day. Polyurethane foam cubes containing the inoculum from the 5 l BTF were added to 118 ml serum bottles (following the protocol described in Chapter 6 section 6.2.3). The batches were incubated for 42 days with 5 or 100% $^{13}$C-labeled CH$_4$ ($^{13}$CH$_4$). The AOM occurrence was assessed by analyzing the gas stable isotope composition of CH$_4$ and CO$_2$ by gas chromatography-isotope ratio mass spectrometry (GC-IRMS), as explained in Chapter 6 (section 6.2.4). The stable isotope composition ($\delta^{13}$C, Chapter 6) of CO$_2$ during the incubation increased, showing that CH$_4$ was converted to CO$_2$ (Figure 7.3b). However, the AOM activity was ~10 times lower for the incubation with 100% $^{13}$CH$_4$ than with 5% $^{13}$CH$_4$ (Figure 7.3b), showing that the microorganisms involved in AOM (probably ANME) had difficulties metabolizing heavy CH$_4$ (100% $^{13}$CH$_4$), as was shown previously (Milucka et al., 2012; Scheller et al., 2016).

Five samples for each type of incubation were withdrawn at different time intervals (0, 21, 28, 35 and 42 days) and fixed in 1% paraformaldehyde (fixation described in Chapter 6, section 6.2.6). The samples were placed on conductive surface polycarbonate filters after embedding and sectioning. The cells were hybridized by CARD-FISH with specific targeting probes for the identification and quantification of ANME-2 and DSS (following the CARD-FISH protocol in Chapter 6, section 6.2.6). The target cells were visualized with an epifluorescence microscope and marked with a laser microdissection system (LMD) (Figure 7.3c). The samples were successfully hybridized with the DSS probe (DSS658) and ANME-2 probe (ANME-2 538), separately. The cells hybridized with the DSS probe were abundant (~80%) (Figure 7.3c). Only few cells were hybridized with the ANME-2 probe (< 5%) and therefore only the results on the DSS cells are shown here.

Four samples hybridized with the DSS probe were chosen for NanoSIMS analysis: unlabelled, 5% $^{13}$CH$_4$ withdrawn on days 0 and 42 and 100% $^{13}$CH$_4$ on day 42 (Figure 7.3b). Mainly C, S and N isotopes in single cells were detected and localized in 1 or 2 LMD-marked spots. Secondary ion images of $^{12}$C, $^{13}$C, $^{12}$C$^{14}$N, $^{13}$C$^{14}$N, $^{32}$S, $^{31}$P, $^{19}$F were recorded simultaneously by the secondary mass serial quantitative secondary ion mass spectrometer. Figure 7.3d shows the micrograph of a target spot taken by epifluorescence microscopy after CARD-FISH compared to the micrograph acquired by NanoSIMS. The Look@NanoSIMS programme (Polerecky et al., 2012) was used to analyze the NanoSIMS data and individual isotopic ratios for each single cell were determined. The regions of interests (ROIs) were defined by comparing a CARD-FISH image with the respective NanoSIMS acquired image. The CARD-FISH images were used to identify the single DSS cells and the elemental and isotopic compositions of each ROI were exported as graphical and text-based formats.

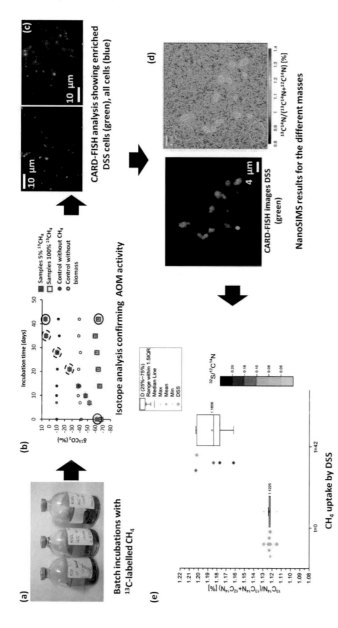

**Figure 7.3** Preliminary results obtained by fluorescence in-situ hybridization–nanometer scale secondary ion mass spectrometry (FISH-NanoSIMS). (a) Image of batches incubatated with polyurethane foam cubes and biomass from the BTF and $^{13}CH_4$. (b) $CO_2$ stable isotope composition ($\delta^{13}CO_2$) along the labeling incubation experiment (42 days) for samples incubated with 5 or 100% $^{13}CH_4$ and controls with nitrogen and without biomass. The black circles indicate the time points from which the NanoSIMS analysis was done, while the dashed circles show the time points to be analyzed. (c) CARD-FISH images of microbial cells enriched at the end of the BTF operation. DAPI stained cells (blue) and DSS (green). (d) CARD-FISH image with DSS of the spot targeted with NanoSIMS and image showing the $100 \times (^{13}C^{14}N/(^{12}C^{14}N + ^{13}C^{14}N))$ in a $50 \times 50$ μm field of analysis for sample at day 0 incubated with 5% $^{13}CH_4$. (e) $100 \times (^{13}C^{14}N/(^{12}C^{14}N + ^{13}C^{14}N))$ for each DSS cell at day 0 and 42 for sample incubated with 5% $^{13}CH_4$. Each DSS cell is colored upon the abundance of $^{32}S$.

Mainly, two different ratios were calculated for each ROI and class: $^{13}C^{14}N$ ratio, as $^{13}C^{14}N/(^{12}C^{14}N+^{13}C^{14}N)$ and $^{32}S/^{12}C^{14}N$ to determine the enrichment in $^{13}C$ and the abundance of $^{32}S$ (Milucka et al., 2012; Polerecky et al., 2012). Using 5% $^{13}CH_4$, the DSS cells had a mean $^{13}C^{14}N$ ratio of 1.12 and 1.18 on days 0 and 42, respectively (Figure 7.3e). The $^{13}C^{14}N$ enrichment in the DSS cells between days 0 and 42 was measurable, but it only increased by 0.06 % (Figure 7.3e). The $^{32}S/^{12}C^{14}N$ calculated for each ROI defined also the abundance of sulfur within the cells. After 42 days of incubation with 5% $^{13}CH_4$, the cells contained more sulfur than those observed on day 0 (Figure 7.3e).

The increase in $^{13}C^{14}N$ within the DSS, despite being small, shows the possible uptake of $CH_4$ by the targeted DSS. However, considering these results and the bulk isotope measurements obtained by GC-IRMS (Figure 7.3b), more hybridized and marked samples need to be analyzed by NanoSIMS for cellular level investigation of DSS. As the AOM activity slowed down in the last period of incubation (between days 38 and 40, Figure 7.3e), it is necessary to analyze the samples incubated with 5% $^{13}CH_4$ on day 28 and 35. Moreover, the ANME-2 cells should also be analyzed to fully understand and quantify the carbon fluxes within the microorganisms fed by $CH_4$ and sulfate.

In Chapters 5 and 6, ANME and SRB acclimated to deep sediment conditions were enriched in BTFs at ambient pressure and temperature. The BTF, operated for 230 days and using sulfate as electron acceptor for AOM, had a SR rate of 0.3 mmol l$^{-1}$ day$^{-1}$ , which is half of the highest volumetric rate (0.6 mmol l$^{-1}$ day$^{-1}$) obtained in a membrane bioreactor operated at atmospheric pressure for 884 days (Meulepas et al., 2009a). However, in our study, sulfate was immediately consumed and the highest rate was achieved after 140 days (Chapter 6), which shows that the BTF reduced the usually long start-up time of AOM-SR. Moreover, in the BTF using thiosulfate as electron acceptor, the sulfide production (0.5 mmol l$^{-1}$ day$^{-1}$) and SR (0.4 mmol l$^{-1}$ day$^{-1}$) rates were higher than using sulfate (0.2 mmol l$^{-1}$ day$^{-1}$ and 0.3 mmol l$^{-1}$ day$^{-1}$, respectively). Therefore, the initial addition of thiosulfate in the BTF enhanced the SR activity.

The high porosity of the polyurethane foam used as the packing material in the BTF, offered good biomass retention capacity and enhanced the gas to liquid mass transfer of the poorly soluble $CH_4$ by increasing gas-liquid mixing and retaining $CH_4$ within the pores of the polyurethane foam (Aoki et al., 2014; Cassarini et al., 2017; Estrada et al., 2014b) (Figure 7.4). Therefore, the BTF commonly used for aerobic/anoxic waste gas treatment is also a suitable bioreactor configuration for the enrichment of slow growing microorganisms, such as ANME and SRB, at ambient pressure and temperature using deep marine sediments as the inoculum. Moreover, the BTF may be more effective than membrane reactors for accelerating microbial growth in the start-up period and it reduces the operational costs of high pressure reactors operating at ambient conditions and enhancing methane bioavailability by good methane retention within the polyurethane foam.

## 7.4    Future perspectives

Based on the good results obtained in this study, it is noteworthy to mention that a BTF is a suitable bioreactor configuration for the enrichment of ANME and SRB using marine sediments as the inoculum. Moreover, the sediment from the marine Lake Grevelingen is a suitable inoculum, as it showed the highest AOM rates at low pressure conditions (0.1 and 0.45 MPa), despite the scarce $CH_4$ availability. As depicted in Figure 7.4, thiosulfate as the electron acceptor for AOM can be used to activate the sediment for AOM-SR and enrich the SRB community. In full-scale operations, in order to obtain high AOM-SR rates and enrich both ANME and SRB in a BTF, sulfate should be used as the sole electron acceptor, as ANME were not enriched with thiosulfate as electron acceptor. This research, thus, highlights the possibility of applying a new strategy for environmental bioremediation applications brings the attention into further investigation on the role of the microbes involved and their metabolic activities.

One of the most significant aspect that requires further investigation relates to the proposed syntrophic association between ANME and SRB, which is still under debate. If such syntrophy occurs through direct electron transfer as proposed by McGlynn et al. (2015) and Wegener et al. (2015), it is necessary to understand the role of the proteins responsible of the electron transfer and the function of nanowires observed. Moreover, growing ANME separately, without the bacterial partner, would be advantageous to fully understand their metabolism. Recently, it has been shown that ANME can be decoupled from the bacterial partner using an external electron acceptor (Scheller et al., 2016). These microorganisms are apperently capable of exporting electron outside the cell and it would be interesting investigating if they can be electronically conductive (McGlynn, 2017). In order to achieve this, bio-electrochemical systems (BES, Chapter 2 section 2.5.5) could be used to study the electron transfer mechanisms and enrich the ANME separately.

As reported in Chapter 2, there are different techniques that can be used to study metabolic activities and functions of the microorganism (e.g. metagenomic analysis), however, such analysis require highly enriched ANME communities which are very difficult to obtain. An efficiently designed bioreactor for the enrichment of the AOM community can be a scientific breakthrough for the further exploration of the ecophysiology of ANME and its potential biotechnological application. Thus, enriching ANME is important and a BTF was suggested in this study for their enrichment at ambient conditions.

However, different reactions took place in the BTFs showing difficulties to fully control the AOM-SR process and optimizing the system for SR. The bioprocess could be controlled by a combination of continuous monitoring of the products and mathematical modeling. For instance, sulfide and pH can be continuously monitored by using pH and pS (sulfide sensor) electrodes, so that the sulfide can be removed before reaching the toxic threshold. Besides, different sulfur compounds (e.g. elemental sulfur and polysulfides) were formed in the BTFs, which are difficult to quantify, limiting the understanding of the role sulfur and of the processes involved in the BTFs. Therefore, new methods for quantitative analysis of elemental sulfur and polysulfides in solid and liquid phases need to be further investigated.

Moreover, a process control algorithm can be developed for the SR process (Cassidy et al., 2015). The use of such control systems will improve the long term performance of the BTF, regulate the growth conditions of the different microorganisms and optimize the dosing levels of different electron donors and acceptors.

The BTF technology is widely used for the treatment of industrial waste gases containing volatile organic and inorganic pollutants, showing high removal efficiencies of pollutants, at low concentrations and high gas flow rates (Guerrero & Bevilaqua, 2015; Kennes & Veiga, 2013; Niu et al., 2014). Besides, the operating and capital costs of treating pollutants using a BTF are usually low compared to other physico-chemical approaches (Mudlar et al. 2010). However, accumulation of excess biomass and clogging has been often reported as the main disadvantages of using this technology for waste gas treatment. Differently, the BTFs used in this study for AOM-SR did not pose any operational problem such as clogging or channeling, and the biomass was actively maintained during its long term operation (>200 days). In this study, we proposed polyurethane foam as packing material and the BTF was operated in sequential fed-batch mode for the trickling liquid-phase, while the gas-phase $CH_4$ was continuously supplied to the BTF in up-flow mode. Nevertheless, the BTF design for a slow metabolic process such as AOM-SR can be further optimized.

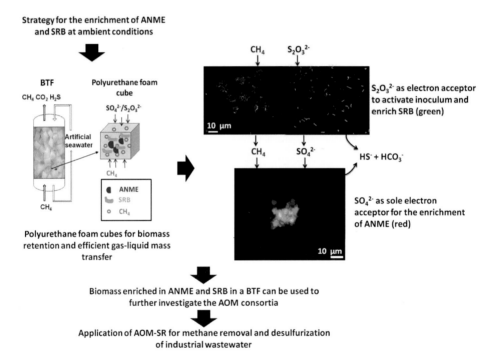

**Figure 7.4** Strategy for the enrichment of ANME and SRB at ambient conditions in a biotrickling filter and future applications.

Previous reports have shown that the BTF has also been used to treat $CH_4$ emissions under aerobic conditions (Estrada et al., 2014a; Estrada et al., 2014c). Estrada et al. (2014b) showed that, using a gas-recycling strategy, the $CH_4$ to liquid-phase mass transfer can be enhanced even at low concentrations of $CH_4$. In the BTF for AOM-SR, recycling the non-oxidized $CH_4$ to achieve complete removal might be a good approach. However, unwanted or toxic compounds need to be stripped out prior recirculation. For instance, the hydrogen sulfide produced can be continuously removed by integrating other bioprocesses for achieving metal precipitation (i.e. $ZnCl_2$) to avoid inhibition of the microbial growth.

Besides polyurethane foam, different packing materials have also been successfully used in BTF for wastegas treatment. These materials are either organic or inorganic, such as, molecular sieves, ceramic rings, compost, coconut fiber, activated carbon, stones and resins (Avalos et al. 2012; Chen et al. 2016; Mudlar et al. 2010). For the enrichment of ANME and SRB, inert materials are preferred and polyurethane foam was chosen as it resembles the carbonate chimneys, which are often present in natural ANME habitats such as carbonate nodules (Marlow et al., 2014). For future applications, naturally occurring materials (e.g. sandstone, lava rocks) or inert materials such as plastic rings or resins can be tested in a BTF for AOM-SR.

Based on the knowledge gained from this work, ANME and SRB can be enriched in a BTF. The enriched community can be further used to understand its mechanisms and the BTF design can be further improved and controlled for future biotechnology applications of AOM-SR.

## 7.5    References

Aoki, M., Ehara, M., Saito, Y., Yoshioka, H., Miyazaki, M., Saito, Y., Miyashita, A., Kawakami, S., Yamaguchi, T., Ohashi, A., Nunoura, T., Takai, K., Imachi, H. 2014. A long-term cultivation of an anaerobic methane-oxidizing microbial community from deep-sea methane-seep sediment using a continuous-flow bioreactor. *PLoS ONE*, **9**(8), e105356.

Avalos Ramirez, A., Jones, J.P., Heitz, M. 2012. Methane treatment in biotrickling filters packed with inert materials in presence of a non-ionic surfactant. *J. Chem. Technol. Biotechnol.*, **87**(6), 848-853.

Bhattarai, S., Cassarini, C., Naangmenyele, Z., Rene, E.R., Gonzalez-Gil, G., Esposito, G., Lens, P.N.L. 2017. Microbial sulfate-reducing activities in anoxic sediment from marine lake Grevelingen: screening of electron donors and acceptors. *Limnology*, **19**(1), 31-41.

Cassarini C., Rene E. R., Bhattarai S., Esposito G., Lens P.N.L. (2017) Anaerobic oxidation of methane coupled to thisoulfate reduction in a biotrickling filter. *Bioresour. Technol.*, **240**(3), 214-222.

Chen, Y., Wang, X., He, S., Zhu, S., Shen, S. 2016. The performance of a two-layer biotrickling filter filled with new mixed packing materials for the removal of $H_2S$ from air. *J. Environ. Manage.*, **165**(8), 11-16.

Deusner, C., Meyer, V., Ferdelman, T. 2009. High-pressure systems for gas-phase free continuous incubation of enriched marine microbial communities performing anaerobic oxidation of methane. *Biotechnol. Bioeng.*, **105**(3), 524-533.

Estrada, J.M., Dudek, A., Muñoz, R., Quijano, G. 2014a. Fundamental study on gas-liquid mass transfer in a biotrickling filter packed with polyurethane foam. *J. Chem. Technol. Biotechnol.*, **89**(9), 1419-1424.

Estrada, J.M., Lebrero, R., Quijano, G., Pérez, R., Figueroa-González, I., García-Encina, P.A., Muñoz, R., 2014b. Methane abatement in a gas-recycling biotrickling filter: evaluating innovative operational strategies to overcome mass transfer limitations. *Chem. Eng. J.*, **253**(53), 385-393.

Finster, K. 2008. Microbiological disproportionation of inorganic sulfur compounds. *J. Sulfur Chem.*, **29**(3-4), 281-292.

Guerrero, R.B., Bevilaqua, D. 2015. Biotrickling filtration of biogas produced from the wastewater treatment plant of a brewery. *J. Environ. Eng.*, **141**(8), 04015010.

Kennes, C., Veiga, M.C. 2013. Bioreactors for waste gas treatment. in: Kennes, C., Veiga, M.C. (Eds.), (Vol 4), Springer Netherlands, Dodrecht, the Netherlands.

Krüger, M., Treude, T., Wolters, H., Nauhaus, K., Boetius, A. 2005. Microbial methane turnover in different marine habitats. *Palaeogeogr. Palaeocl.*, **227**(1-3), 6-17.

Losekann, T., Knittel, K., Nadalig, T., Fuchs, B., Niemann, H., Boetius, A., Amann, R. 2007. Diversity and abundance of aerobic and anaerobic methane oxidizers at the Haakon Mosby mud volcano, Barents Sea. *Appl. Environ. Microbiol.*, **73**(10), 3348-3362.

Marlow, J.J., Steele, J.A., Ziebis, W., Thurber, A.R., Levin, L.A., Orphan, V.J. 2014. Carbonate-hosted methanotrophy represents an unrecognized methane sink in the deep sea. *Nat. Commun.*, **5**(5094), 1-12.

McGlynn, S.E. 2017. Energy metabolism during anaerobic methane oxidation in ANME archaea. *Microbes Environ.*, **32**(1), 5-13.

Meulepas, R.J.W., Jagersma, C.G., Gieteling, J., Buisman, C.J.N., Stams, A.J.M., Lens, P.N.L. 2009. Enrichment of anaerobic methanotrophs in sulfate-reducing membrane bioreactors. *Biotechnol. Bioeng.*, **104**(3), 458-470.

Milucka, J., Ferdelman, T.G., Polerecky, L., Franzke, D., Wegener, G., Schmid, M., Lieberwirth, I., Wagner, M., Widdel, F., Kuypers, M.M.M. 2012. Zero-valent sulphur is a key intermediate in marine methane oxidation. *Nature*, **491**(7425), 541-546.

Musat, N., Halm, H., Winterholler, B., Hoppe, P., Peduzzi, S., Hillion, F., Horreard, F., Amann, R., Jørgensen, B.B., Kuypers, M.M.M. 2008. A single-cell view on the ecophysiology of anaerobic phototrophic bacteria. *Proc. Natl. Acad. Sci. USA*, **105**(46), 17861-17866.

Musat, N., Musat, F., Weber, P.K., Pett-Ridge, J. 2016. Tracking microbial interactions with NanoSIMS. *Curr. Opin. Biotech.*, **41**(7), 114-121.

Nauhaus, K., Boetius, A., Kruger, M., Widdel, F. 2002. *In vitro* demonstration of anaerobic oxidation of methane coupled to sulphate reduction in sediment from a marine gas hydrate area. *Environ. Microbiol.*, **4**(5), 296-305.

Niemann, H., Duarte, J., Hensen, C., Omoregie, E., Magalhaes, V.H., Elvert, M., Pinheiro, L.M., Kopf, A., Boetius, A. 2006a. Microbial methane turnover at mud volcanoes of the Gulf of Cadiz. *Geochim. Cosmochim. Ac.*, **70**(21), 5336-5355.

Niemann, H., Losekann, T., de Beer, D., Elvert, M., Nadalig, T., Knittel, K., Amann, R., Sauter, E.J., Schluter, M., Klages, M., Foucher, J.P., Boetius, A. 2006b. Novel microbial communities of the Haakon Mosby mud volcano and their role as a methane sink. *Nature*, **443**(7113), 854-858.

Niu, H., Leung, D.Y., Wong, C., Zhang, T., Chan, M., Leung, F.C. 2014. Nitric oxide removal by wastewater bacteria in a biotrickling filter. *J. Environ. Sci.*, **26**(3), 555-565.

Polerecky, L., Adam, B., Milucka, J., Musat, N., Vagner, T., Kuypers, M.M. 2012. Look@ NanoSIMS–a tool for the analysis of nanoSIMS data in environmental microbiology. *Environ. Microbiol.*, **14**(4), 1009-1023.

Scheller, S., Yu, H., Chadwick, G.L., McGlynn, S.E., Orphan, V.J. 2016. Artificial electron acceptors decouple archaeal methane oxidation from sulfate reduction. *Science*, **351**(6274), 703-707.

Schreiber, L., Holler, T., Knittel, K., Meyerdierks, A., Amann, R. 2010. Identification of the dominant sulfate-reducing bacterial partner of anaerobic methanotrophs of the ANME-2 clade. *Environ. Microbiol.*, **12**(8), 2327-2340.

Suarez-Zuluaga, D.A., Weijma, J., Timmers, P.H.A., Buisman, C.J.N. 2014. High rates of anaerobic oxidation of methane, ethane and propane coupled to thiosulphate reduction. *Environ. Sci. Pollut. Res.*, **22**(5), 3697-3704.

Templer, S.P., Wehrmann, L.M., Zhang, Y., Vasconcelos, C., McKenzie, J.A. 2011. Microbial community composition and biogeochemical processes in cold-water coral carbonate mounds in the Gulf of Cadiz, on the Moroccan margin. *Mar. Geol.*, **282**(1-2), 138-148.

Vigneron, A., Cruaud, P., Pignet, P., Caprais, J.-C., Cambon-Bonavita, M.-A., Godfroy, A., Toffin, L. 2013. Archaeal and anaerobic methane oxidizer communities in the Sonora Margin cold seeps, Guaymas Basin (Gulf of California). *ISME J.*, **7**(8), 1595-1608.

Zhang, Y., Henriet, J.-P., Bursens, J., Boon, N. 2010. Stimulation of *in vitro* anaerobic oxidation of methane rate in a continuous high-pressure bioreactor. *Biores. Technol.*, **101**(9), 3132-3138.

# Appendix 1

# Supporting Information for Chapter 4

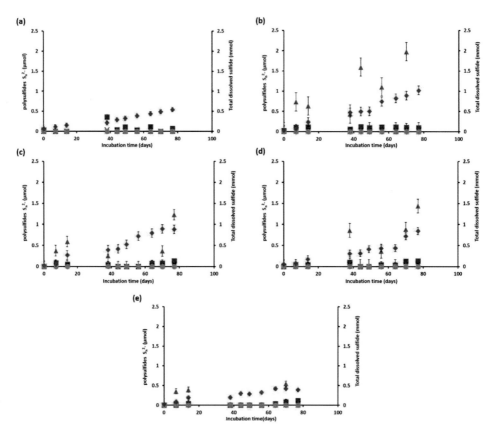

**Figure S4.1**. Total dissolved sulfide (◆) and polysulfides concentration, namely $S_2^{2-}$ (●), $S_3^{2-}$ (✕), $S_4^{2-}$ (✗), $S_5^{2-}$ (■). $S_6^{2-}$ (▲), during the incubation of Grevelingen sediment at (**a**) 0.45 MPa, (**b**) 0.1MPa, (**c**) 10 MPa, (**d**) 20 MPa, and (**e**) 40 MPa. Error bars indicate the standard deviation (n=3).

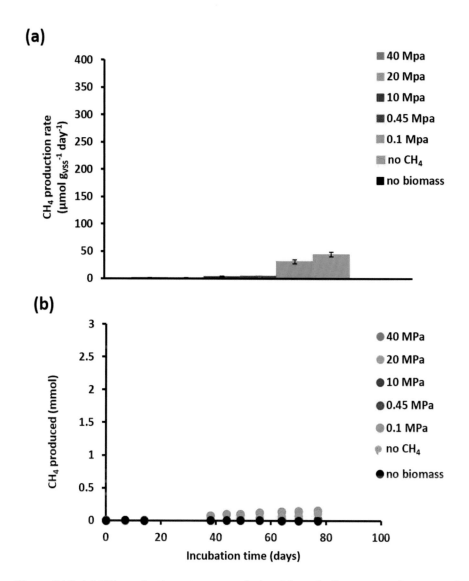

**Figure S4.2**. (a) CH$_4$ production rates were calculated from the linear regression over at least four successive measurements in which the calculated $^{12}$CH$_4$ increase over time was linear. (b) The CH$_4$ produced was calculated from the $^{12}$CH$_4$. Methanogenic activity and CH$_4$ produced during AOM were determined for incubations at different pressures and controls without CH$_4$, but with N$_2$ in the headspace and without biomass. Error bars indicate the standard deviation (n=4).

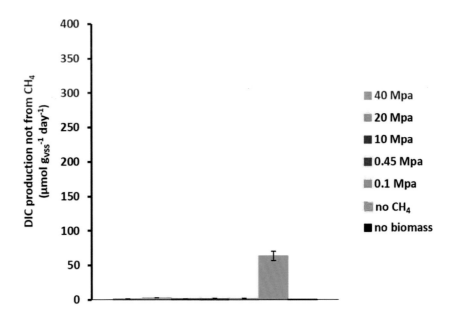

**Figure S4.3** Dissolved inorganic carbon (DIC) production rate not due to methanotrophy calculated from $^{12}CO_2$ produced for incubation at different pressure and controls without $CH_4$ but with $N_2$ in the headspace and without biomass. Error bar indicates the standard deviation (n=3).

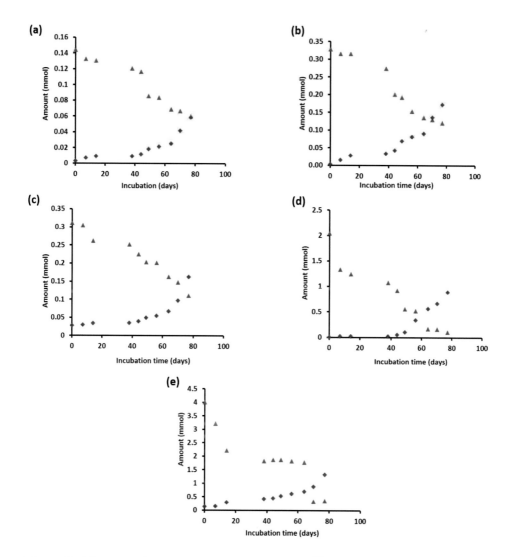

**Figure S4.4** Concentration profiles of methane oxidized ($^{13}CH_4$, ▲) and dissolved inorganic carbon (DIC, ◆) calculated from the produced $^{13}CO_2$) for the incubation of Grevelingen sediment at (a) 0.1 MPa, (b) 0.45 MPa, (c) 10 MPa, (d) 20 MPa and (e) 40 MPa.

## Biography

Chiara Cassarini was born on 21$^{st}$ January 1987 in Bologna, Italy. Chiara completed her bachelor (BSc) in 2010 in Industrial Chemistry at the University of Bologna (Italy) and she obtained her MSc degree in Geochemistry at Utrecht University (the Netherlands). She collaborated closely with the Deltares Company in the Netherlands and with the department of Isotope Biogeochemistry in Leipzig (UFZ) Germany where she used compound-specific stable isotope analysis (CSIA) to investigate the degradation of chlorinated compounds in an industrial contaminated site. Simultaneously, during her MSc studies, she participated in a research cruise on the North Sea to investigate ocean acidification. Chiara got admitted into the Erasmus Mundus Joint Doctorate Program on Environmental Technologies for Contaminated Solids, Soils and Sediment (ETeCoS$^3$). She started as a PhD fellow at UNESCO-IHE (Delft, the Netherlands) from October 2013, but the research was conducted in other partner institutions as required by the Doctorate programme. Chiara worked with experts from different scientific groups, such as the UFZ in Leipzig, Jiao Tong University in Shanghai (China) and the University of Cassino and Southern Lazio (Italy). Within the framework of her PhD studies, she investigated how various sulfur compounds, as electron acceptors, impact the anaerobic oxidation of methane in packed bed systems. She used various approaches including, CSIA, FISH and FISH-NanoSIMS.

## Publications and conferences

### Publications

Cassarini C., Rene, E.R., Bhattarai, S., Esposito, G., Lens, P.N.L. (2017) Anaerobic oxidation of methane coupled to thiosulfate reduction in a biotrickling filter. *Bioresour. Technol.*, **240**(3), 214-222.

Bhattarai, S.*, **Cassarini, C.***, Naangmenyele, Z., Rene, E. R., Gonzalez-Gil, G., Esposito, G., Lens, P.N.L. (2017) Microbial sulfate reducing activities in anoxic sediment from marine lake Grevelingen. *Limnology*, **19**(1), 31-41. (*Both authors have contributed equally to this paper)

Bhattarai, S., **Cassarini, C.**, Naangmenyele, Z., Rene, E.R., Gonzalez-Gil, G., Esposito, G., Lens, P.N.L. (2017) Anaerobic methane-oxidizing microbial community in a coastal marine sediment: anaerobic methanotrophy dominated by ANME-3. *Microb. Ecol.* **74**(3), 608-622.

Bhattarai, S., **Cassarini, C.**, Rene, E. R., Kümmel, S., Esposito, G. and Lens, P. N. L. (2018) Enrichment of ANME-2 dominated anaerobic methanotrophy from cold seep sediment in an external ultrafiltration membrane bioreactor. *Eng. Life Sci.* doi:10.1002/elsc.201700148.

Bhattarai, S., **Cassarini, C.**, Rene, E. R., Zhang, Y., Esposito, G. and Lens, P. N. L. (2018) Enrichment of sulfate reducing anaerobic methane oxidizing community dominated by ANME-1 from Ginsburg Mud Volcano (Gulf of Cadiz) sediment in a biotrickling filter. *Bioresour. Technol.* Doi: 10.1016/j.biortech.2018.03.018.

### Conferences

**Cassarini, C.**, Gil-Gonzalez, G., Rene, E.R., Vogt, C., Musat, N., Lens, P.N.L. Anaerobic methane oxidation by different sulfur compounds as electron acceptors in a bioreactor. Poster Presentation at EMBO Workshop on Microbial Sulfur Metabolism, Helsingør (Denmark), April 2015.

**Cassarini**, C., Rene, E.R., Zhang, Y., Vogt, C., Musat, N., Esposito, G., Lens, P.N.L. Enrichment of microorganisms involved in anaerobic oxidation of methane in bioreactors, 3rd International Conference on Biogas Microbiology (ICBM-3), Wageningen, Netherlands, May, 2017.

Netherlands Research School for the
Socio-Economic and Natural Sciences of the Environment

# D I P L O M A

### For specialised PhD training

The Netherlands Research School for the
Socio-Economic and Natural Sciences of the Environment
(SENSE) declares that

## *Chiara Cassarini*

born on 21 January 1987 in Bologna, Italy

has successfully fulfilled all requirements of the
Educational Programme of SENSE.

Gaeta, 28 June 2017

the Chairman of the SENSE board

Prof. dr. Huub Rijnaarts

the SENSE Director of Education

Dr. Ad van Dommelen

K O N I N K L I J K E   N E D E R L A N D S E
A K A D E M I E   V A N   W E T E N S C H A P P E N

The SENSE Research School declares that Ms Chiara Cassarini has successfully fulfilled all requirements of the Educational PhD Programme of SENSE with a work load of 44.5 EC, including the following activities:

SENSE PhD Courses

o   Environmental research in context (2014)
o   Research in context activity: 'Government review of the IPCC AR5 Synthesis Report' (2015 )

Other PhD and Advanced MSc Courses

o   Environmental Engineering, IHE-Delft (2014)
o   Environmental Technologies for Contaminated solids, soils and sediments (ETeCoS3) introductory course, Cassino University (UNICAS) (2014)
o   ETeCoS3 summer school on biological treatment of solid waste, University of Cassino (2014)
o   ETeCoS3 Summer School on Contaminated Soils, University of Paris-Est (2015)
o   ETeCoS3 Summer School on Contaminated Sediments - Characterization and Remediation, IHE-Delft (2016)

External training at a foreign research institute

o   Course on NanoSIMS analysis and poster presentation, Technical University of Munich, Germany (2014)
o   Training on different microscopy techniques, Helmholtz Centre for Environmental Research-UFZ, Leipzig, Germany (2015)

Management and Didactic Skills Training

o   Assisting MSc course 'Microbiology in Urban Water and Sanitation ' (2013-2014)
o   Supervising two MSc students 'Anaerobic oxidation of methane and sulphate reduction by marine sediments in the presence of different electron acceptors and electron donors' (2014) and 'Biological sulphate removal from wastewater using anaerobic sludge and enriched marine sediment' (2014-2015)

Oral Presentations (a.o.)

o   *Anaerobic methane oxidation and the microorganisms involved: enrichment, kinetics and morphology.* PhD Symposium From Water Scarcity to Water Security, 29-30 September 2016, Delft, The Netherlands
o   *Enrichment of microorganisms involved in anaerobic oxidation of methane in bioreactors.* 3rd International Conference on Biogas Microbiology (ICBM-3), 1-3 May 2017, Wageningen, The Netherlands

SENSE Coordinator PhD Education

Dr. Peter Vermeulen

For Product Safety Concerns and Information please contact our EU
representative GPSR@taylorandfrancis.com Taylor & Francis Verlag GmbH,
Kaufingerstraße 24, 80331 München, Germany

Printed and bound by CPI Group (UK) Ltd, Croydon, CR0 4YY
02/05/2025
01859932-0002